Hemangiomas and Vascular Malformations of the Head and Neck

Hemangiomas and Vascular Malformations of the Head and Neck

Edited by
Milton Waner, M.D.
and
James Y. Suen, M.D.

A JOHN WILEY & SONS, INC., PUBLICATION

New York • Chichester • Weinheim • Brisbane • Singapore • Toronto

This text is printed on acid-free paper. ⊗

Published simultaneously in Canada.

For ordering and customer service, call 1-800-CALL-WILEY.

Library of Congress Cataloging in Publication Data:
Hemangiomas and vascular malformations of the head and neck / edited
    by Milton Waner and James Y. Suen.
        p.   cm.
    Includes bibliographical references and index.
    ISBN 0-471-17597-8 (cloth : alk. paper)
    1. Birthmarks. 2. Hemangiomas. 3. Head—Blood-vessels—
Abnormalities. 4. Neck—Blood-vessels—Abnormalities. I. Waner,
Milton. II. Suen, James Y., 1940-
    [DNLM: 1. Head and Neck Neoplasms—therapy. 2. Hemangioma—
therapy. 3. Blood Vessels—abnormalities. WE 707 H487 1998]
RL793.H46   1998
616.99'315—dc21
DNLM/DLC
for Library of Congress   98-10332
                                                           CIP

Printed in the United States of America

10 9 8 7 6 5 4 3 2 1

Contents

3 The Natural History of Vascular Malformations   47
MILTON WANER, M.D. AND JAMES Y. SUEN, M.D.

4 The Diagnosis of a Vascular Birthmark   83
MILTON WANER, M.D. AND JAMES Y. SUEN, M.D.

5 The Surgical Pathology Approach to Pediatric Vascular Tumors and Anomalies   93
PAULA E. NORTH, M.D. AND MARTIN C. MIHM., JR., M.D.

8 Treatment Options for the Management of Hemangiomas   233

MILTON WANER, M.D. AND JAMES Y. SUEN, M.D.

9 The Treatment of Hemangiomas   263

MILTON WANER, M.D. AND JAMES Y. SUEN, M.D.

Contributors

JOSEPH BROGDEN, M.A.
Behavior Medicine Consultant
Vascular Anomalies Center
Arkansas Children's Hospital
800 Marshall Street
Little Rock, AR 72202

CHARLES A. JAMES, M.D.
Assistant Professor
Department of Radiology
University of Arkansas for Medical Sciences
Arkansas Children's Hospital
800 Marshall Street
Little Rock, AR 72202

MARTIN C. MIHM, JR., M.D., F.A.C.P.
Clinical Professor of Pathology
Harvard University Medical School
Boston, MA
Senior Dermatopathologist
Massachusetts General Hospital
Boston, MA

PAULA E. NORTH, M.D., PH.D.
Assistant Professor
Department of Pathology
University of Arkansas for Medical Sciences
Arkansas Children's Hospital
800 Marshall Street
Little Rock, AR 72202

James Y. Suen, M.D., F.A.C.S.
Professor and Chairman
Department of Otolaryngology—Head and Neck Surgery
University of Arkansas for Medical Sciences
800 Marshall Street
Little Rock, AR 72202

Milton Waner, M.D., F.C.S. (S.A.)
Director, Vascular Anomalies Center
Arkansas Children's Hospital
Professor
Department of Otolaryngology—Head and Neck Surgery
University of Arkansas for Medical Sciences
800 Marshall Street
Little Rock, AR 72202

Preface

Congenital vascular anomalies have been, and to many, still remain among the most poorly understood entities. Misdiagnosis and mismanagement are still too often encountered and "Modern Medicine" seems to have left these unfortunate patients in the Dark Ages. A renaissance began with the "biological" classification proposed by Mulliken and Glowacki in 1982. This was indeed a giant step forward. For the first time, we were able to both accurately diagnose congenital vascular lesions and understand their natural history. Certain dogmas still persisted however; chief of which was the notion that all hemangiomas involuted completely and therefore required no treatment. After seeing many of the sequelae of hemangiomas and the psychological trauma from them, we feel that intervention is indicated in many cases. Based on the knowledge that hemangiomas do involute, the policy of "benign neglect" that has been the standard for many years deprives patients of recent technological advances which are now available.

Over the last decade, a ground swell of physicians has begun to challenge this blanket policy of benign neglect. Modern medicine has indeed progressed to where we are able to offer many of these patients an alternative. But, given the confines and limitations of our various specialties, many physicians looked for a solution within their own field of expertise and attempted to apply it to all cases. As a consequence, several useful modalities exist, all of which have a role in the management of congenital vascular lesions, but none of which are appropriate for all lesions at all times. A fresh look at all of the modalities and their usefulness was therefore essential. This text was born out of that notion and we believe advocates each modality where it is most suited. Evidence-based medicine teaches us of the importance of data from randomized therapeutic trials, but many conditions are not suitable for such trials. There are no clear end points that occur in a relatively short period of time in this field, and for this reason there have been no randomized prospective clinical trials. We must therefore exercise our best judgment and not lose sight of our objectives. We hope that this text will be helpful to those in the various specialties who are faced with these complex problems.

MILTON WANER, M.D.
JAMES Y. SUEN, M.D.

Little Rock, Arkansas
January 1999

Acknowledgments

Two individuals have had a profound influence on my career and therefore indirectly on this book. The first, Bruce N. P. Benjamin, taught me the analytical skills essential to a good clinician. Dr. Benjamin is without doubt one of the finest teachers from whom I have had the privilege of learning. Although he may not fully realize the extent of his influence, he has helped shape the clinical skills of many otolaryngologists. His profoundly logical approach to a problem is one I have always tried to follow. The next is my co-author, James Y. Suen. Dr. Suen is the most brilliant, meticulous surgeon from whom I have ever learned. His skill, techniques, and attention to detail are responsible for many of the procedures we are now able to perform. Needless to say, without his influence, many of these procedures would not have been possible. I would also like to acknowledge my sons, Daniel and Jarrod, my daughter, Alexandra, and my loving wife, Janee, all of whom made many sacrifices during the preparation of this manuscript. Last but by no means least, my special thanks to Shawn Morton of John Wiley & Sons, who recognized the need for this text and despite the many deadlines that came and went, never appeared to lose his patience. Lisa Van Horn also deserves a special mention. Her tireless production editing made the final text something of which to be proud.

M.W.

When I first met Milton Waner, I was impressed with his knowledge and his ability to make the complicated seem simple. Through the years, this continues to be his strong point. Dr. Waner is extremely bright, upbeat, thoughtful, and innovative. His mind is always in over-drive. In many ways he fits the description of a genius. I am proud of his accomplishments and of our friendship. This book is his project, and he deserves all of the credit.

J. Y. S.

Hemangiomas
and Vascular
Malformations
of the Head and Neck

A Classification of Congenital Vascular Lesions

MILTON WANER, M.D., F.C.S. (S.A.) AND
JAMES Y. SUEN, M.D., F.A.C.S.

Prior to the early 1980s, the study of congenital vascular lesions was in a state of confusion. Despite several attempts, the lack of a uniformly accepted classification and of a clear understanding of their natural history were in part to blame. This was further confounded by the misconception that most of these lesions would spontaneously disappear within the first few years of life. Congenital vascular lesions were therefore, in the main, misdiagnosed and left untreated. If we, the physicians, were confused, what of our patients? The plight of many of these unfortunate persons led them from physician to physician in search of answers that, all too often, left them even more confused.

Hemangiomas and Vascular Malformations of the Head and Neck, Edited by Waner, M.D. and Suen, M.D.
ISBN 0471-17597-8 © 1999 Wiley-Liss, Inc.

The renaissance began in the early 1980s. Mulliken and Glowacki (1982) introduced a "biological" classification that finally dispensed with the old, confusing terminology and replaced it with a set of clear and concise terms and definitions. Based on clinical and biological differences, they recognized two distinct entities: hemangiomas and vascular malformations. Hemangiomas are usually *not* present at birth, *proliferate* during the first year of life, and then *involute.* The endothelial cells, responsible for this proliferation, exhibit a high rate of turnover during this period. Vascular malformations, on the other hand, are *always* present at birth, *never proliferate,* and *never involute.* Their rate of endothelial cell turnover is thus always normal. Unlike previous attempts, this classification was unique in its approach. Its basis was clinical, and it was easily applied. Once a diagnosis had been made, one was able to appreciate the natural history of the lesion and was thus better equipped to plan treatment. All the guess work and ambiguity had finally disappeared. This classification has become widely accepted and forms the basis for this text. In the natural course of events, Mulliken and Glowacki's work has evolved and some minor changes have been made (see Table 1.1). These are all, none theless, in keeping with the fundamental bases of the "biological" classification.

TABLE **1.1**
Mulliken and Glowacki's Classification of Congenital Vascular Lesions (Modified by Waner and Suen)

Hemangiomas
Vascular malformations
Venous
Venular
Capillary
Lymphatic
Arteriovenous
Mixed
Venous–lymphatic
Venous–venular

1.1. HEMANGIOMAS

Hemangiomas first appear soon after birth, although as many as 30% may already be apparent at birth (Mulliken, 1988). They typically proliferate during the first year of life and then involute during the childhood years. Although considerable variation is seen with respect to their rate of proliferation and involution, these two features, proliferation and involution, are constant (Mulliken and Glowacki, 1982). Furthermore, it is these same two features that distinguish hemangiomas from vascular malformations, which neither proliferate nor involute.

Although hemangiomas can and occasionally do present in most organs, the vast majority are found in or just deep to skin. Hemangiomas that originate in the papillary dermis usually stretch the overlying skin as they proliferate and present as a bright red macular or papular mass. These lesions were referred to as "strawberry" or "capillary hemangiomas" (Fig. 1.1). A lesion originating within the

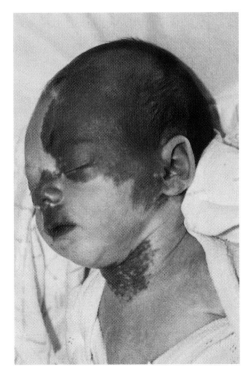

Figure **1.1**
A "capillary" or, more appropriately, a superficial hemangioma.

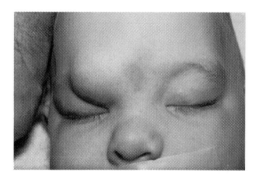

Figure 1.2
A "cavernous" or, more appropriately, a deep hemangioma.

reticular dermis or subcutaneous tissue may also expand the overlying skin, but, because the lesion is deeper and thus separated from the epidermis by a layer of collagen (the papillary dermis), it will impart a bluish hue to the epidermis or appear colorless (Fig. 1.2). This type of lesion was previously known as a "cavernous hemangioma." On occasion, both superficial and deep components will be present, and the resultant lesion will have elements of both a capillary hemangioma and a cavernous hemangioma. These lesions were known as "capillary cavernous hemangiomas" (Fig. 1.3). Since all hemangiomas constitute a single entity regardless of their depth with respect to skin, terms such as *capillary* and *cavernous* are misleading and should be discarded in favor of more meaningful nomenclature. Strawberry or capillary hemangiomas are in reality superficial *(superficial hemangiomas)* and cavernous hemangiomas are deep *(deep hemangiomas)*. The unfortunate term *capillary cavernous hemangiomas* is an oxymoron and should be replaced with the term *compound hemangioma* (Table 1.2).

During the proliferative phase, tubules of plump proliferating endothelial cells with frequent mitoses are the dominant histological features (Mulliken and Glowacki, 1982). As the end of proliferation draws near, mast cells become more abundant, these plump proliferating endothelial cells become less active, and, as this happens, they progressively flatten until, during involution, hemangiomas are made up of flat, inactive, normal-appearing endothelial cells, surrounding large ectatic vascular channels in a matrix of fibro-fatty tissue (Mulliken and Glowacki, 1982). Involuting hemangiomas are therefore not dissimilar in histological appearance from vascular malformations, and, were it not for their distinctive behavior, it would be difficult to tell them apart.

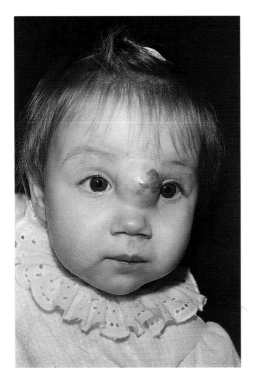

Figure 1.3
A "capillary cavernous" or, more appropriately, a compound hemangioma.

1.2. VASCULAR MALFORMATIONS

Vascular malformations are always present at birth, although they may not be apparent and only present at a later stage (Mulliken and Glowacki, 1982). In contradistinction to hemangiomas, vascular malformations do not proliferate nor do they involute. Instead, slow but relentless expansion throughout the course of the patient's life is the

TABLE 1.2
Hemangiomas: Old Versus New Nomenclature

Old	New
Capillary or strawberry hemangioma	Superficial hemangioma
Cavernous hemangioma	Deep hemangioma
Capillary cavernous hemangioma	Compound hemangioma

norm (Barsky et al., 1980). Intercurrent trauma, sepsis, hormonal modulation, and changes in blood or lymph flow and/or pressure may,
however, temporarily speed up this process. Progressive ectasia of the
existing vascular structures and not vascular proliferation is responsible for this expansion. The diameters of the constituent vessels thus increase, but not their number. Vascular malformations therefore expand
by a process of hypertrophy (enlargement due to an increase in size of
existing structures). Hemangiomas, on the other hand, enlarge by a
process of hyperplasia (enlargement due to an increase in cell number).
Throughout the life of a vascular malformation, flat endothelial cells
with a normal rate of turnover surrounding malformed and/or ectatic
vessels are the characteristic histological features (Mulliken and
Glowacki, 1982).

Vascular malformations may be further classified according to the
type of vessel they are made of. Mulliken and Glowacki (1982) recognized arteriovenous, capillary, venous, lymphatic malformations, and
mixed malformations. Over the course of several years and after an extensive histological survey of a large number of vascular malformations, we have made minor changes in these subcategories (see Chapter 4 and Table 1.1). The fundamental basis of the classification,
however, remains unchanged. The first of these changes concerns arteriovenous malformations. Recent evidence suggests that their underlying abnormality is at the level of the capillary bed across which the arteriovenous shunting takes place (Baker et al., 1998). All other findings,
including hypertrophy of the afferent arterial system and dilation of
the efferent venous system, are secondary effects resulting from the altered flow dynamics across this primary abnormality. Baker et al. (1998)
hypothesize that an absence or lack of innervation of the precapillary
sphincters within the capillary bed is responsible for this unimpeded
shunting from the arterial system across the capillary bed to the venous
system. The nidus of the malformation is thus the abnormal capillary
bed, and these are more appropriately renamed *capillary malformations.*
However, in order to avoid confusion, we will use the term arteriovenous malformation throughout the text.

The second change concerns port-wine stains. According to
Mulliken and Glowacki, port-wine stains are capillary malformations,
yet histological analysis confirms the fact that they are made up of
ectatic postcapillary venules in the papillary plexis (Barsky et al., 1980).
The term *venular malformation* is thus more accurate and will be used.

1.3. CLINICAL APPLICATION

Not only have Mulliken and Glowacki provided us with some semblence of order in what was once a morass of confusion, their classification also allows us to establish a diagnosis clinically, without the necessity or the expense of special investigations and/or hazardous biopsies. On arriving at a diagnosis, one is then able to confidently predict the natural history of the lesion and thus plan management in a more relevant way. Since all hemangiomas eventually involute, one can safely explain this to the child's parents and, when appropriate, treat the child with this in mind. On the other hand, a vascular malformation will never involute; because in most cases progressive expansion with time can be expected, one may elect to intervene at an earlier stage.

The true brilliance of this classification lies in its simplicity. In the vast majority of cases an accurate history and a thorough clinical examination are all that is needed. With regard to the history, two questions are of paramount importance (Table 1.3): (1) When was the lesion first noticed? (2) Has it grown since then?

A lesion that is present at birth can be either a hemangioma or a vascular malformation, but only a hemangioma will have proliferated. If, however, the child is seen at an early stage in the cycle of a hemangioma, or prior to proliferation, physical examination or repeated evaluation will be necessary to establish whether the lesion is proliferating.

TABLE **1.3**
An Approach to the Diagnosis of a Vascular Lesion

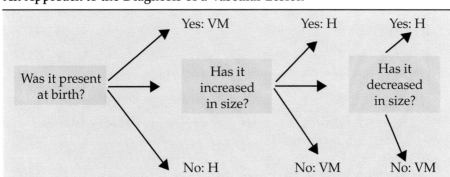

While this is somewhat of a simplification, the answers represent the most likely diagnosis. H, hemangioma; VM, vascular malformation.

A lesion not present at birth but that proliferates is almost certainly a hemangioma. Vascular malformations may, however, be present at birth but not noticed for days, weeks, months, or, in the case of an arteriovenous malformation, years after birth. These lesions never proliferate but do increase in size through hypertrophy of existing structures (Table 1.4). This is, as a rule, slow but progressive and extends over a longer period of time. On occasion, parents are unsure whether the lesion was present at birth. The second question will then, in most cases, help with the diagnosis. When confronted with an older child for the first time, a third question concerning whether the lesion has diminished in size is helpful. Clearly, only hemangiomas will shrink or involute. Unfortunately, lymphatic malformations may fluctuate in size with upper respiratory tract infections and this may be confusing. With regard to expansion, a clear association with infections will distinguish these lesions from hemangiomas. Lymphatic malformations expand during infections, whereas hemangiomas do not.

TABLE **1.4**

A Comparison of Hemangiomas and Vascular Malformations

Hemangiomas	Vascular Malformations
Clinical Features	
Presents neonatal or early infancy	Always present at birth
Proliferates during first year of life	Never proliferates
Involutes after first year of life	Never involutes
M:F ratio is 1:6	M:F ratio is 1:1
Biological Features	
Proliferation: Plump, proliferating endothelial cells, numerous mast cells	Flat, normal endothelial cells
Involution: Apoptosis and progessive flattening of endothelial ectatic cells. Larger, more obvious but fewer vascular channels, with perivascular fibro-fatty tissue	Normal endothelial turnover; malformed, vascular channels: venules, veins, capillaries or hymphatic vessels

In most cases, the diagnosis will be obvious on physical examination. Should the diagnosis still be equivocal, repeated evaluations over the ensuing 3–4 months will most likely establish the diagnosis. Bright red papular lesions are characteristic findings of a superficial hemangioma. Bright red macular lesions may either be a port-wine stain (vascular malformation) or an early hemangioma (Fig. 1.4). Port-wine stains are, however, in most instances uniform, whereas hemangiomas usually appear stippled or telangiectatic. A bluish mass may be a subcutaneous hemangioma, a venous malformation, or an arteriovenous malformation (Table 1.5; Fig. 1.5). In this instance, palpation will be helpful. Proliferating hemangiomas and arteriovenous malformations feel firm and "rubbery," whereas venous malformations are much softer and compressible. They can, in most cases, completely empty with compression. This is extremely unusual in the case of a hemangioma or an arteriovenous malformation. A deep hemangioma may also be confused with a lymphatic malformation that had not been apparent at birth (Fig. 1.6). Lymphatic malformations, however, are softer and can be transilluminated, whereas hemangiomas are much firmer and do not transilluminate. Again, if one is still unsure of the diagnosis after a thorough examination, parental reassurance and subsequent

TABLE **1.5**
Differential Diagnosis of a Bluish Mass

Subcutaneous hemangioma
    Firm, rubbery
        Definite history of proliferation
        Definite history of involution
Venous malformation
    Soft, compressible
        Empties with compression
        Expands with raised venous pressure
Arteriovenous malformation
    Firm
        Fluid thrill
        Dilated tortuous vessels

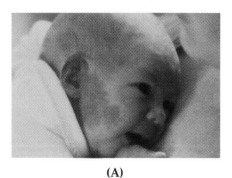
(A)

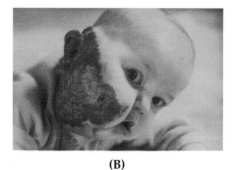
(B)

Figure 1.4
(A) A bright red macular lesion. This could be either a venular malformation (port-wine stain) or an early hemangioma. **(B)** A subsequent examination revealed obvious proliferation, making the diagnosis of hemangioma obvious.

re-evaluation will often be helpful. A hemangioma is bound to declare itself by proliferating during the first year of life. Lastly, a sarcoma or some other fast-growing soft tissue tumor may mimic a vascular lesion (Fig. 1.7). This is extremely uncommon but can usually be resolved with magnetic resonance imaging.

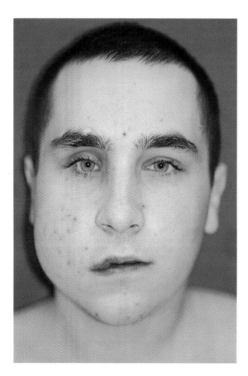

Figure 1.5
A bluish mass. This lesion could have been a subcutaneous hemangioma, an arteriovenous malformation, or a venous malformation. Subsequent examination revealed proliferation. This was therefore a hemangioma. A detailed history revealed that the mass was present at birth and has steadily increased in size. It expands when the patient is recumbent. This is therefore likely to be a venous malformation.

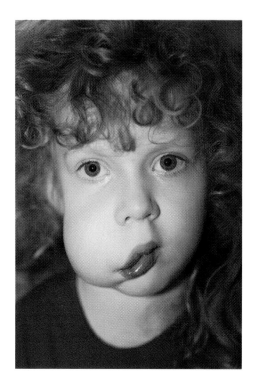

Figure 1.6
A lymphatic malformation of the buccal fat space. The lesion fluctuated in size with upper respiratory infections.

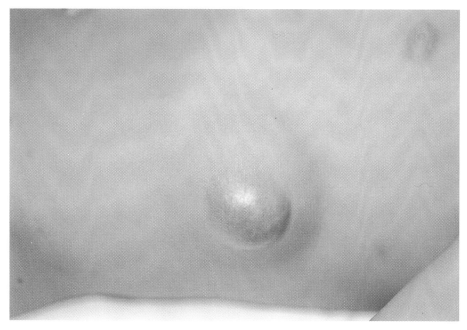

Figure 1.7
A "vascular mass" initially diagnosed as an arteriorvenous malformation turned out to be a soft tissue tumor. A preoperative magnetic resonance imaging scan would have excluded the diagnosis of an arteriovenous malformation.

REFERENCES

Baker, L., Waner, M., Thomas, R., Suen, JY.: The management of arteriovenous malformations. Arch. Otolaryngol., 1998.

Barsky, S.H., Rosen, S., Greer, D.E., and Noe, J.M.: The nature and involution of port-wine stains: A completely assisted study. J. Invest. Dermatol. 74:154, 1980.

Mulliken, J.B.: Diagnosis and natural history of hemangiomas. In *Vascular Birthmarks: Hemangiomas and Vascular Malformations.* Philadelphia: W.B. Saunders, 1988.

Mulliken, J.B., Glowacki, J.: Hemangiomas and vascular malformations in infants and children: A classification based on endothelial characteristics. Plast. Reconstr. Surg. 69:412, 1982.

BIBLIOGRAPHY

Bivings, L.: Spontaneous regression of angiomas in children: Twenty-two years of observation covering 236 cases. J. Pediatr. 45:643, 1954.

Bowers, R.E., Graham, E.A., and Tomlinson, K.M.: The natural history of the strawberry nevus. Arch. Dermatol. 82:667, 1960.

Fraser, J.: The hemangioma group of endothelioblastomata. Br. J. Surg. 7:335, 1919–1920.

Lister, W.A.: The natural history of strawberry naevi. Lancet 1:1429, 1938.

Margileth, A.M., and Museles, M.: Cutaneous hemangiomas in children: Diagnosis and conservative management. JAMA 194:135, 1965.

Simpson, J.R.: Natural history of cavernous haemangiomata. Lancet 2:1057, 1959.

Stigmar, G., Crawford, J.S., Ward, C.M., and Thomson, H.G.: Ophthalmic sequelae of infantile hemangiomas of the eyelids and orbit. Am. J. Ophthalmol. 85:805, 1978.

The Natural History of Hemangiomas

MILTON WANER, M.D., F.C.S. (S.A.) AND
JAMES Y. SUEN, M.D., F.A.C.S.

Hemangiomas are the most common tumors in infants. Approximately 10%–12% of all white children and, curiously, 22% of preterm infants weighing less than 1,000 g are affected (Holmdahl, 1955; Jacobs, 1957; Amir et al., 1986). The incidence among Asian and black children appears to be much lower, 0.8%–1.4% (Hidano and Nakajima, 1972; Pratt, 1967). These figures should, however, be viewed with caution because their source predates the current classification and may well have included vascular malformations in the respective series. Although hemangiomas are not believed to be familial, 10% of affected infants had a positive family history (Margileth and Museles, 1965). Furthermore, given that 10% of all white infants are affected, it seems logical that 1 out of every 10 of these children will have a positive

Hemangiomas and Vascular Malformations of the Head and Neck, Edited by Waner, M.D. and Suen, M.D.
ISBN 0471-17597-8 © 1999 Wiley-Liss, Inc.

family history. In addition, a recent study of twins with hemangiomas failed to implicate hereditory factors in the etiology of the hemangiomas; a familial basis therefore seems unlikely (Cheung et al., 1997).

Although 30% of hemangiomas are present at birth, the majority present during the first few weeks of life. The earliest sign of an impending hemangioma is a blanched macule (Figs. 2.1, 2.2). This is frequently not noticed and is soon followed by the development of discrete telangiectasia, often surrounded by a blanched halo (Fig. 2.3) (Hidano and Nakajima, 1972; Payne et al., 1966). The contrast between the blanched area and normal tissue will be especially obvious when the child cries. Although the seemingly paradoxical appearance of a blanched macule as the earliest sign of a nascent proliferative vascular lesion defies logic, its occurrence is really quite reasonable. As the endothelial cells become active, they "plump up" and, by so doing, temporarily block the internal vascular space, leaving little or no room for red blood cells, hence the blanched appearance. Continued proliferation will eventually increase the vessel diameter, thereby allowing more room for whole blood and thus eliminate the blanched appearance. As the process continues, an increase in vessel number will give rise to the appearance of discrete telangiectasia, which, by sheer force of vessel number, eventually coalesces to form a red macule.

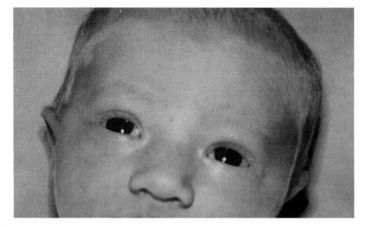

Figure 2.1
An infant with a blanched macule, a precurser to a hemangioma, between her eyebrows.

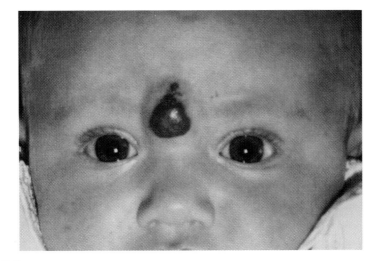

Figure 2.2
The same infant 10 weeks later, with an obvious hemangioma occupying the same spot.

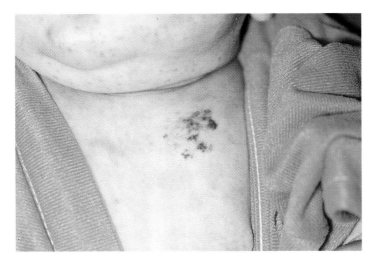

Figure 2.3
An infant with a blanched macule on her anterior chest wall and signs of early telangiectasia developing in the center of the macule.

Most hemangiomas (80%) are solitary focal lesions, and the head and neck is the most common site (60%) (Margileth and Museles, 1965; Finn et al., 1983). Recent work has shown that there are definite sites of predilection (Waner et al., 1998). In the head and neck, focal lesions may appear anywhere along a line starting in the middle of the cheek and passing lateral to the eyebrow, then over the top of the eyebrow to the glabella. From there the line passes medial to the medial canthus, down the paranasal area, and then across the alar groove to the nasal tip and the columella. The midline of the upper lip and the lateral aspect of the lower lip are also among the most frequently involved sites (Fig. 2.4). Although the reason for these sites is not known, they appear to correspond with lines of embryological fusion. Perhaps, totipotential cells at these sites retain their ability to develop into endothelial cells and/or pericytes, and, under the influence of angiogenesis stimulation, they develop into blood vessels. A less common, alternative pattern of diffuse involvement tends to follow the facial dermatone distribution (Fig. 2.5).

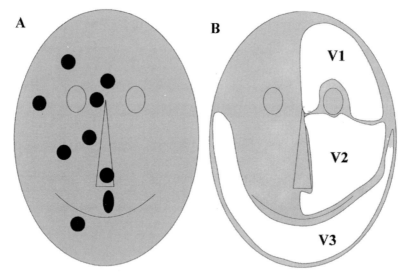

Figure 2.4
(A) A schematic diagram showing some of the sites of predilection of focal hemangiomas of the face. (B) An alternative pattern with diffuse involvement tends to follow trigeminal (V) dermatomes.

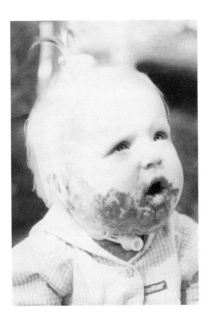

Figure 2.5
A 6-month-old child with a diffuse hemangioma involving the V3 dermatome. This pattern of involvement is often associated with ulceration and airway obstruction at the level of the hypopharynx.

2.1. PROLIFERATION

Hemangiomas characteristically proliferate during the first year of life. In rare circumstances, a hemangioma may be fully grown at birth and then proceed to involute rapidly within a few months of life (Boon et al., 1996). Although the rate and timing of growth within this first year is extremely variable, two periods of rapid growth are frequently seen (Table 2.1). The first and most characteristic is during the neonatal period and in early infancy (Mulliken, 1988). Following this, a second growth spurt is commonly seen between 4 and 6 months of age. The ultimate size of the lesion depends on the degree and the duration of proliferation. Unfortunately, there is no way of predicting this (Figs. 2.6–2.8).

Proliferating hemangiomas are characterized by plump proliferating endothelial cells (Mulliken and Glowacki, 1982). In the early stages, sheets of proliferating endothelial cells are evident with apparent disorganization and an absence of vascular channels (Fig. 2.9). Reticulin staining, however, demonstrates that there is organization in that the proliferation is limited by a basement membrane. The cells are so

TABLE 2.1
A Schematic Representation of the Rate of Proliferation of a Hemangioma

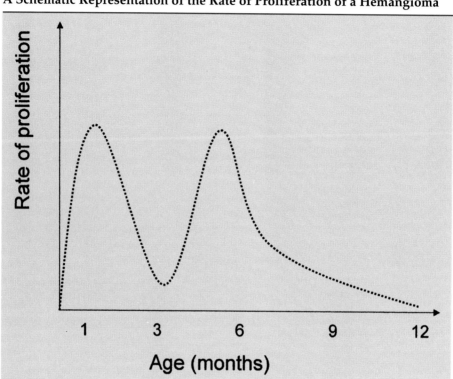

plump that they take up the entire vascular space. As proliferation progresses, more organization becomes apparent. Obvious vascular channels become evident, and blood components become clearly visible within their lumens. Toward the end of proliferation, the hemangioma becomes organized into lobules separated by fibrous septae, each with its own blood supply and venous drainage. Occasionally these feeding vessels can become quite impressive, especially in the presence of a large hemangioma. The role of mast cells appears to be central to this process. Kessler et al. (1976) observed that mast cells accumulated in tumors prior to the ingrowth of new capillaries. It was later discovered that heparin, released by the mast cells, facilitated the migration of endothelial cells toward the tumor (Azizkhan et al., 1980). Mast cells are also reported to be present in proliferating hemangiomas in extremely high concentrations (Pasyk et al., 1983; Mulliken and Glowacki, 1982). Recent work appears to have shed more light on the

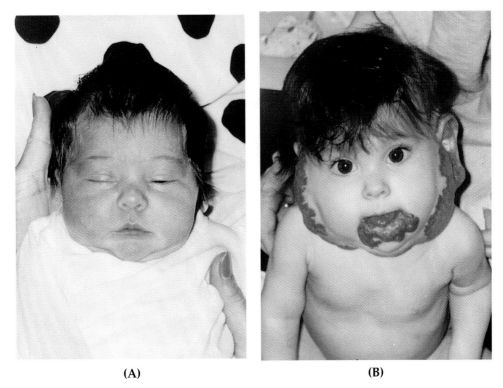

(A)   (B)

Figure 2.6
(A) A 2-day-old infant. (B) The same child 12 weeks later with an obvious hemangioma in the distribution of the mandibular branch of the trigeminal nerve.

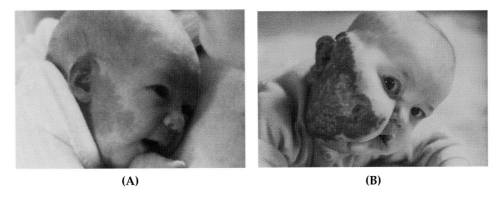

(A)   (B)

Figure 2.7
(A) An infant with an erythematous macule overlying his parotid area. (B) The same infant 12 weeks later with a massive hemangioma involving his parotid gland.

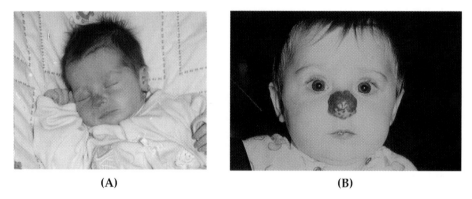

(A)   (B)

Figure 2.8
(A) An infant with an erythematous macule involving her nasal tip. (B) The same child several months later with an obvious compound nasal tip hemangioma.

role of mast cells. Takahashi et al. (1994) found that mast cells occurred in their highest concentrations during the early and middle involuting phases, (i.e., between 13 months and 60 months) and not as previously reported (Fig. 2.10). This work corroborates earlier mast cell counts by Pasyk et al. in 1983. From an analysis of histological and immunohisto-chemical data, Takahashi and coworkers (1994) postulate that during

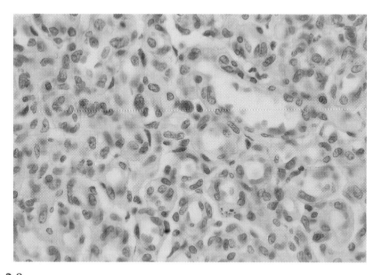

Figure 2.9
A section of a proliferating hemangioma (hematoxylin and eosin stain [H&E]) showing the plump proliferating endothelial cells in apparent disorganization. Few vascular lumina are evident.

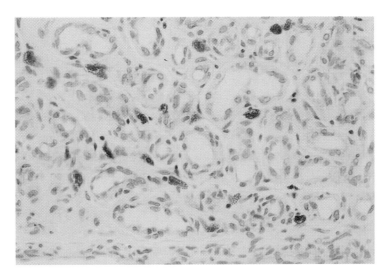

Figure 2.10
A toluidine blue stain of a mature hemangioma at the end of its proliferative phase. Note the presence of numerous mast cells.

the third trimester immature endothelial cells co-exist with immature pericytes. These cells maintain their proliferative capacity for a limited period during postnatal life. Rapid proliferation of these cells, mediated in part by angiogenic peptides such as β-fibroblast growth factor-β and proliferating cell nuclear antigen, marks the beginning of a clinically apparent hemangioma. These peptides may also induce the differentiation of these cells, which in turn signal an influx of mast cells as well as the induction of tissue inhibitors of metalloproteinases (TIMPs). TIMPs together with modulators expressed by the mast cells (interferon and transforming growth factor) stop the proliferation of endothelial cells and thereby induce involution. Thus, Takahashi et al. (1994) believe that the process of involution is passive with senescence of endothelial cells, fibrosis, and the formation of fat. Mihm and North (1997) believe that apoptosis of the endothelial cells is primarily responsible for involution and that this is therefore an active process. (personal communication, 1997)

Proliferating endothelial cells within a hemangioma rest on a multilaminated basement membrane. Multilamination is also seen in vasculitides such as scleroderma and Raynaud's disease and is therefore believed to be a characteristic of cyclical endothelial proliferation and death.

The physical features of the hemangioma will depend on its depth with respect to skin, its size, and its stage in involution. A superficial proliferating hemangioma will present as a bright red mass, whereas a deep lesion will be bluish or colorless, the main difference being the depth of the lesion with respect to the collagen layer of the papillary dermis (Figs. 1.1, 1.2.). Because collagen scatters visible light and red light penetrates deeper than blue light, red blood vessels appear bluish through collagen in the same way that a deep nevus, made of dark brown pigment, will appear blue through collagen. During involution superficial lesions will turn a more dusky color, and, by late involution, they will have assumed a decidedly purple color (Fig. 2.11). With regard to size, there is great variation. Hemangiomas range in size from a pinhead to a massive lesion almost the size of the infant's head (Figs. 2.12, 2.13). The consistency of the lesion will also depend on its stage of evolution. Proliferating hemangiomas will have a firm consistency and will expand with raised intravascular pressure. During active proliferation, a rapidly growing lesion will be extremely tense and, contrary to current opinion, we believe they are painful.

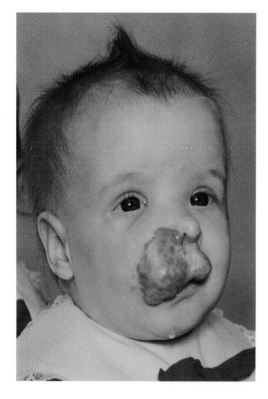

Figure 2.11
A mature hemangioma in its early stages of involution. Note the dusky purplish appearance of the surface component.

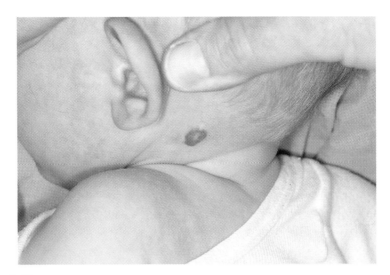

Figure 2.12
A small occipital hemangioma.

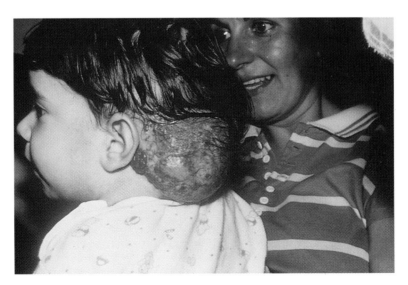

Figure 2.13
An infant with a massive occipital hemangioma. The difference between this and
the hemangioma shown in Figure 2.12 is the size of the lesion. Both are occipital
hemangiomas and probably have their origins in the same site.

2.2. INVOLUTION

Proliferation invariably slows down, and by the end of the first year of life involution begins. These states are not distinct, but overlap considerably. The transition from proliferation to involution is gradual and seems to coincide with the appearance of mast cells and TIMPs (Takahashi et al., 1994). Histological analysis suggests that islands of decreased endothelial activity first make their appearance at this stage. These islands slowly and relentlessly expand until they become the dominant feature, and the scales are tipped in favor of involution. There has been much speculation as to the cause of involution. Several authors have thought that thrombosis was partly responsible, although no histological evidence to support this theory has surfaced. Histological studies have shown that as the proliferating plump endothelial cells become less active, they progressively flatten so that in late involution flat, inactive endothelial cells predominate (Figs. 2.14, 2.15). At the same time, the vascular channels become more obvious until large ectatic capillary-like vessels are seen (Mulliken and Glowacki, 1982). The hemangioma has thus changed from a predomi-

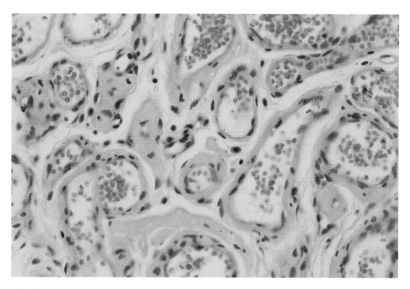

Figure 2.14
An H&E section of a hemangioma in its early stage of involution. Note the flatting of the endothelial cells and the presence of very few mitoses among these cells. Vascular channels are now obvious.

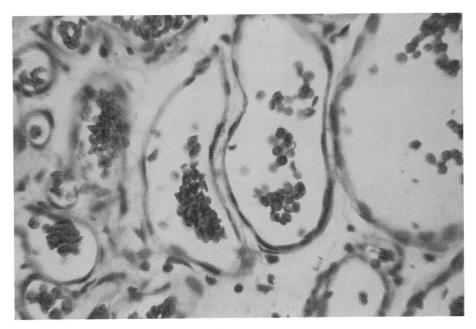

Figure **2.15**
A hemangioma its late phase of involution. Note the flat, normal-appearing endothelial cells and the presence of a fibro-areolar matrix.

nantly cellular to a predominantly vascular lesion. Involution is also characterized by a progressive deposition of perivascular fibro-fatty tissue together with a decrease in the number of vascular channels (Mulliken and Glowacki, 1982). As the number of vascular channels diminishes, the remaining vessels become progressively more ectatic. These ectatic vessels tend to persist in a completely involuted lesion as telangiectasia in a dense collagen reticular framework with islands of fatty tissue.

Clinical evidence of involution commences with a definite diminution in the rate of growth. In time, the hemangioma will become less tender to palpation and will feel less tense. Astute parents will notice that the lesion no longer expands when the child cries, and a cutaneous hemangioma will change its color from bright red to a darker, duskier maroon. Eventually, the lesion will soften and begin to shrink from its center to its periphery in a radial pattern (Fig. 2.16). These signs will obviously be more subtle and hence more difficult to appreciate with deep hemangiomas. Nonetheless, the timing and progression is the same as that seen with cutaneous lesions.

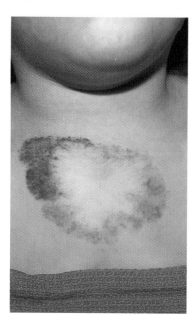

Figure **2.16**
A hemangioma in its late phase of involution.
Note that involution has proceeded from the
center of the lesion to the periphery.

The rate of involution is extremely variable. To date, no known
characteristics appear to influence either the rate of involution or its degree of completion (Finn et al., 1983; Simpson, 1959; Bowers et al.,
1960). Bowers et al. (1960) concluded that 50% of lesions will have involuted by 5 years of age and that a further 20% will have completed this
process by age 7. In the remainder, complete involution may take a further 3–5 years (Bowers et al., 1960; Lister, 1938; Simpson, 1959).

An issue often inadequately dealt with, but an extremely important
one nonetheless, concerns the true meaning of the term *involution*.
Unfortunately, a large proportion of parents have been led to believe
that the hemangioma will disappear completely and leave no trace.
Perhaps the current dogma concerning nonintervention, coupled with
a lack of substantial experience on the part of physicians, is partly to
blame. Lister in 1938 defined *complete disappearance with no trace* as
meaning just that, or, more commonly, "knowing where to look, one
can just discern a few very faint ecstasies, or a stippled scar resembling
a faint vaccination mark." Bowers et al. (1960) defined resolved as
"gone with no trace or so inconspicuous that a thin dusting with powder would make it invisible." In both cases, the fact that many hemangiomas did not disappear to this extent was underemphasized. Finn et
al. (1983) specifically addressed this issue. They defined an excellent

cosmetic result as one in which "no redundant skin, scar, or telangiectasia" was left. In their analysis of 298 hemangiomas, they recognized
two groups of hemangiomas with respect to involution: those that involuted by 6 years of age and those that did not. Thirty-eight percent of
those involuted by age 6 left a residuum in the form of a scar, redundant skin, or telangiectasia. For those that did not involute by 6 years of
age, the outlook was discouraging. In 80%, a significant cosmetic deformity was left at the end of involution. (Finn et al., 1983). Taking into
consideration the fact that about half the lesions will involute by age 6
(Bowers et al., 1960), computation of these findings will clearly show
that only 40% of hemangiomas involute with an acceptable result.
Therefore, 60% of all patients with hemangiomas will require some
form of corrective surgery (Table 2.2). Our own analysis of children
with hemangiomas that had not involuted by 6 years of age revealed

TABLE 2.2
Depiction of Possible Outcomes in the Life Cycle of an Hemangioma

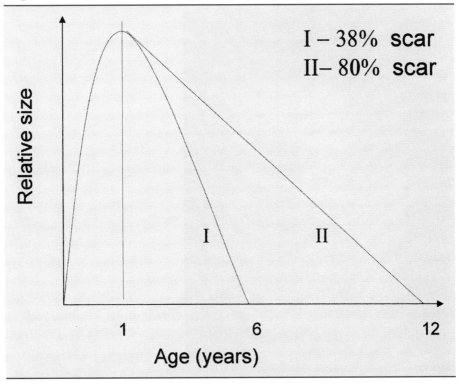

I = Rapid involution; II = Slow involution

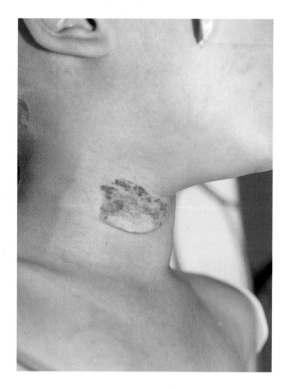

Figure 2.17
The neck of an 8-year-old child showing atrophic scarring with residual fibro-fatty tissue and telangiectasia. These are the residual elements of an involuted hemangioma.

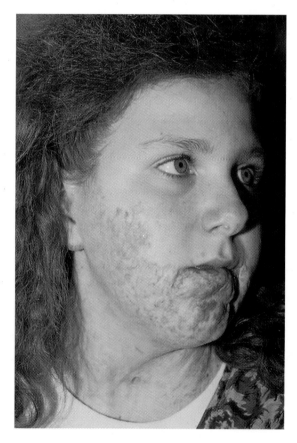

Figure 2.18
A teenager with epidermal atrophy, atrophic scarring, telangiectasia, and residual fibro-fatty tissue. This diffuse V3 dermatome hemangioma had ulcerated during the early stages of its proliferative phase.

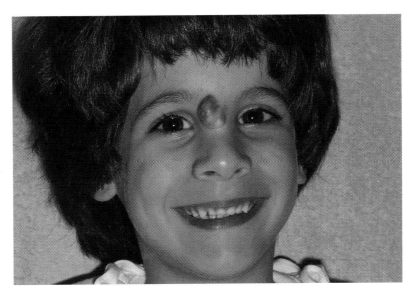

Figure 2.19
A 7-year-old child with residual fibro-fatty tissue, epidermal atrophy and telangiectasia. This hemangioma had not progressed any further and was causing considerable psychosocial trauma.

that cutaneous lesions were likely to leave epidermal atrophy and
telangiectasia, subcutaneous lesions left a deformity due to residual fibro-fatty tissue, and compound hemangiomas left a combination of
epidermal atrophy, telangiectasia, and residual fibro-fatty tissue (Figs.
2.17–2.20).

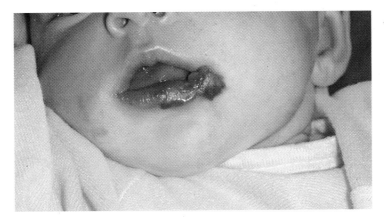

Figure 2.20
An infant with an extensive lower lip hemangioma showing ulceration of the central portion of the hemangioma. If left untreated, the hemangioma will resolve but
the ulcerated area will leave a scar similar to that shown in Figure 5.5.

2.3. COMPLICATIONS OF HEMANGIOMAS

2.3.1. Ulceration

Ulceration is one of the most common complications and occurs in up to 5% of lesion (Margileth and Museles, 1965). Tense, rapidly proliferating hemangiomas as well as hemangiomas in certain anatomical sites such as the upper lip, upper chest, and the anogenital region are more likely to ulcerate than others (Figs. 2.20, 2.21). Two possibilities may explain why hemangiomas ulcerate. Rapid proliferation may distend the overlying skin, which eventually reaches its elastic limit, splits, and, in so doing, forms an ulcer. If proliferation proceeds at a faster rate than the skin can heal (i.e., than the rate of proliferation of the basal cells), ulceration will persist until this situation is reversed. Alternately, because the typical hemangioma derives its blood supply from a circumferential network that in turn gives off feeding branches at right angles to the network, a rapidly proliferating hemangioma may well outstrip its blood supply and necrose at a point most distant from the blood supply (i.e., the center), as is commonly seen (Figs. 2.20, 2.21). Secondary infection is common in lesions that have ulcerated and can

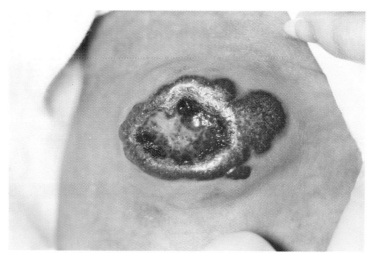

Figure 2.21
An extensive, rapidly proliferating hemangioma showing central necrosis with ulceration.

be difficult to erradicate. Furthermore, hypersensitivity to prolonged use of topical antibiotic ointments is often misdiagnosed as persistent infection and may further compound the problem. Lastly, it must be remembered that lesions that have ulcerated heal by second intension, and this always leaves a residual scar (see Figs. 9.17, 9.23). Parents should be made aware of this as it may alter their decision regarding the timing of intervention. Why subject the child to several years of psychosocial trauma, waiting for an adverse outcome that will most likely need correction?

2.3.2. Airway Obstruction

A hemangioma may rarely obstruct both nasal passages. This could pose a serious problem during the first few weeks of life because these infants are obligate nose breathers. Laryngeal hemangiomas, on the other hand, are much more common and almost always pose a life-threatening problem. The age of onset of symptoms is variable. The vast majority present early, sometime after the first 6–8 weeks of life (which corresponds with the period of rapid proliferation) or, rarely, at a much later stage, when signs of early involution would have been expected (Healy et al., 1980). The reason for this late presentation is not clear. The most common symptoms are inspiratory or biphasic stridor, especially while the infant is feeding or crying. Upper respiratory tract infections also exacerbate symptoms and may lead to the mistaken diagnosis of croup. Cough, cyanosis, and occasionally hoarseness may be present in about 50% of cases and a cutaneous lesion may be present. The diagnosis is usually suspected, especially in the presence of a cutaneous lesion, and confirmed on endoscopy. Additional radiographic procedures that may be helpful include soft tissue penetration (high kilovolt) anteroposterior neck films, which, in the presence of a unilateral lesion, may demonstrate an asymmetrical subglottic narrowing, as might an esophagogram, which is necessary to exclude the possibility of a congenital vascular anomoly (Hollinger et al., 1988; Cooper et al., 1992). A magnetic resonance image should also be considered because it may reveal a large cervical or paratracheal hemangioma adjacent to or continuous with the laryngeal lesion. This is not obvious on endoscopy and will no doubt alter the course of treatment.

The hemangioma may be superficial or deep, localized, or wide-spread. The most common lesion is a superficial, unilateral, subglottic hemangioma (Brodsky et al., 1983). The lesion appears as a firm, red-dish mass and lies just under the mucosal surface. A deeper lesion is usually bluish or whitish and probably lies deep to the conus elasticus or even deep to perichondrium. These lesions are able to extend up the paraglottic space and are thus often more widespread (Fig. 2.22). The term *subglottic* is inappropriate for these lesions; they should probably be known as *laryngeal hemangiomas.* Alternatively, an isolated subglottic lesion may appear as a bluish or whitish mass with no other area of involvement (Fig. 2.23). This is probably an isolated deep (deep to the conus elasticus or perichondrium) lesion.

A large cervical parapharyngeal or palatal hemangioma may also encroach on the upper airway and cause acute or subacute obstruction. This phenomenon appears to be more common with diffuse heman-giomas involving the mandibular dermatome. In the absence of acute obstruction, the insidious onset of obstructive sleep apnea may be pre-sent. This is unfortunately often missed until the child eventually pre-sents with cor pulmonale. The increased work of breathing may also re-sult in failure to thrive (Fig. 2.24).

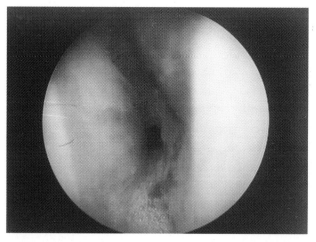

Figure 2.22
A 6-month-old infant with widespread involvement of his right paraglottic space. Note the fullness of the false cord and multiple superficial areas of involvement.

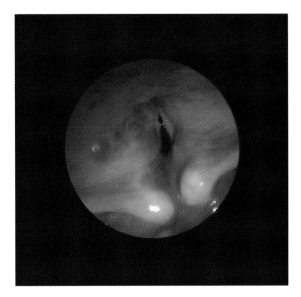

Figure 2.23
An infant with a solitary subglottic hemangioma. The whitish color indicates a
deeper lesion.

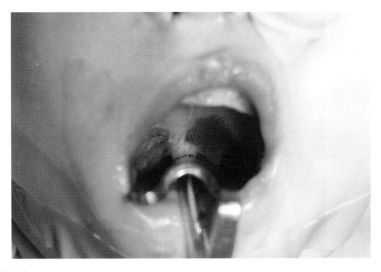

Figure 2.24
An infant with rapidly proliferating palatal hemangioma. The lesion became bulky
and, as a consequence of obstructive sleep apnea, the child failed to thrive.

2.3.3. Auditory Obstruction

Parotid hemangiomas may completely obstruct the external auditory canal and in so doing effect a mild conductive hearing loss. This should not adversely affect the child's development unless this obstruction is bilateral. On the other hand, the accumulation of keratinaceous debris over 2–3 years may result in keratosis obturans. In this instance, aural cleansing should be carried out at 3–6 month intervals or until the obstruction is spontaneously relieved.

2.3.4. Visual Obstruction

Periorbital hemangiomas may obstruct the visual axis and this in turn results in stimulus deprivation amblyopia (Von Noorden, 1974) (Fig. 2.25). Even in the absence of obstruction, anisometric (astigmatic) ambylopia may occur in association with a hemangioma of the upper

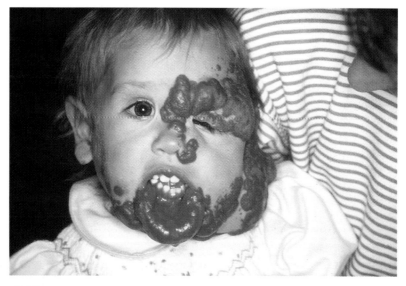

Figure 2.25
An 18-month-old child with multiple cutaneous hemangiomas and obstruction of her visual axis due to an extensive upper eyelid lesion. The child was also in high-output cardiac failure and failed to thrive.

Figure 2.26
An infant with a bulky hemangioma of the upper eyelid. The pressure on the anterior segment of the eye caused astigmatism.

eyelid. This is likely to be a direct consequence of the pressure effect on the anterior segment of the eye (Fig. 2.26) (Robb, 1977). Robb (1977) established this direct cause–effect relationship by determining that the astigmatic axis is related to the location of the hemangioma within the eyelid and that this correlated well with keratometric measurements of corneal astigmatism. Robb (1977) also determined that refractive asymmetry in the form of unilateral myopia was more likely in patients with eyelid hemangiomas, but, as opposed to an astigmatism, no clear correlation with site or size of the lesion could be found. In addition to these complications, both paralytic strabismus and strabismus secondary to amblyopia may be found (Stigmar et al., 1978). Periorbital and orbital hemangiomas may thus cause deprivation amblypoia as a consequence of visual axis obstruction, astigmatism, or myopia.

In spite of the well-documented evidence linking periorbital and orbital hemangiomas with serious sequelae, visual deprivation amblyopia still remains one of the most common causes of preventable blindness (Dunlap, 1971). This has prompted several ophthalmologists to pursue a more aggressive approach to periorbital and/or orbital hemangiomas (Thomson et al., 1979; Kushner et al., 1985; Deans et al., 1992).

2.3.5. Hemorrhage

A frequent concern of parents relates to the bleeding potential of these lesions. Parents are concerned that as the child becomes more mobile, trauma to the lesion, especially in an exposed location, will become more likely to result in severe hemorrhage. While this may occur, it is unlikely, and, as with any other wound, elevation and firm pressure applied for 10–15 minutes should control most bleeding. Ulcerated hemangiomas, on the other hand, are more apt to bleed spontaneously or with minor trauma; although in most of these cases bleeding is not profuse, an occasional massive hemorrhage, severe enough to require transfusion, may occur.

Of more concern is Kasabach-Merritt syndrome (KMS). Profound thrombocytopenia is occasionally seen with a rapidly proliferating solitary hemangioma or multiple cutaneous and systemic lesions and may result in a generalized bleeding disorder (Kasabach and Merritt, 1940; Sutherland and Clark, 1962; Shim, 1968). Affected infants are at risk of pleural, peritoneal, gastrointestinal, and central nervous system hemorrhage. Several factors appear to be responsible for the thrombocytopenia and subsequent hemorrhage. These include an increased rate of sequestration of platelets in the spleen, as well as in the hemangioma itself (Kontras, 1960), and a sequestration of clotting factors within the hemangioma (Margileth and Meuseles, 1965).

KMS is usually seen in the first few weeks of life (the period of rapid proliferation) (Shim, 1968). Rapid expansion of the lesion, edema of the surrounding tissue, petechiae, and ecchymoses are early features of KMS (Fig. 2.27). Later, generalized petechiae and ecchymoses may be seen. Left untreated, a thrombocytopenic hemorrhagic crisis will eventually intervene. It has been stated that KMS is rarely seen in lesions < 5 cm in diameter. Yet edema, ecchymoses, and petechial hemorrhages, all early features of this syndrome, may be seen in infants with lesions considerably less than 5 cm in diameter and, in a few cases, prior to the presence of any palpable mass. KMS is thus more likely to be a function of the type of hemangioma and quite independent of the size of the lesion. Platelet trapping probably occurs very early, but, because the lesion is small at this stage, compensation is more than adequate. However, as the lesion increases in size, compensation will eventually fail to keep up with the losses, manifesting in the full-blown syndrome. Enjolras et al. (1997) believe that all cases of KMS are associ-

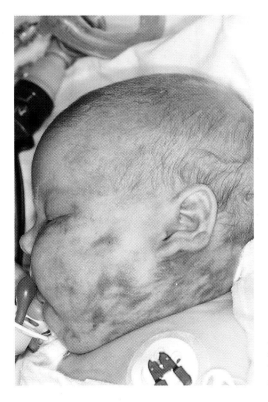

Figure 2.27
An infant with Kasabach-Merritt
syndrome. Note the marked edema
with ecchymoses involving the skin
overlying the lesion.

ated with kaposiform hemangioendothelioma (KHE) and that KMS
does not occur as a complication of classic hemangioma. KHE is indeed
a separate entity and is frequently complicated by KMS. It appears later
in infancy than classic hemangiomas, and its histological features are
distinct from those of hemangiomas. They include infiltrative sheets
and nodules of endothelial cells with islands of dilated lymphatic-type
vessels (Tsang et al., 1991; Zukerberg et al., 1993). The vessels are often
immature and poorly canalized and lack a complete basement membrane. Another constant feature is their being surrounded by a dense
layer of inflammatory tissue (Zukerberg et al., 1993). It therefore seems
likely that all cases of KMS are associated with KHE and not complications of hemangiomas.

The development of disseminated intravascular coagulation (DIC)
may occur together with or independent of KMS. DIC is extremely rare
and has only been reported with massive hemangiomas and in the

presence of sepsis (Ogle et al., 1976; David et al., 1983; Esterly, 1983; Koerper et al., 1983). A consumptive coagulopathy may co-exist with KMS, but this appears to be more an exception than the rule. It is important to recognize the difference between these two syndromes because the use of heparin, commonly used to treat DIC, is contraindicated in KMS as it may potentiate further proliferation of the hemangioma. In an uncomplicated case of KMS, thrombocytopenia can be demonstrated with a simple complete blood count, whereas in the presence of a consumptive coagulopathy the fibrin degradation product level will be elevated and the serum fibrin level will be decreased. It will also be necessary to ascertain the prothrombin time and the partial thromboplastin time.

2.3.6. Congestive Heart Failure

High-output congestive cardiac failure is a potentially lethal complication and carries with it a significant mortality rate. Although a large solitary lesion may be incriminated, high-output cardiac failure is much more commonly seen with multiple cutaneous and/or visceral and/or hepatic lesions (Fig. 2.25) (Cooper and Bolande, 1965; Dupont et al., 1977; Vaksman et al., 1987). Again, the rapidly proliferating phase of growth (early infancy) is the usual stage during which this complication occurs. One of the common ways in which these infants present is with the triad of symptoms of congestive cardiac failure (in the absence of a congenital cardiac abnormality), anemia, and hepatomegaly. Hepatic involvement is almost always accompanied by cardiac failure and may be in the form of multiple hemangiomas or a solitary lesion (Larcher et al., 1981; Pereyra et al., 1982). The resultant hepatomegaly is always out of proportion to the degree of failure. The hemodynamic characteristics of hepatic lesions are unique in that the degree of left to right shunting across these lesions is far greater than one would normally see with a cutaneous lesion of the same size. Although hepatic involvement most often results in cardiac failure, obstructive jaundice, portal hypertension, and occasionally with a large solitary lesion, intestinal obstruction may be the presenting feature (Sademan and Tygstrup, 1974; Helikson et al., 1977; Wishnich, 1978; Larcher et al., 1981).

2.3.7. Skeletal Distortion

Bony distortion is probably more common than the reported incidence of 1% (Boyd et al., 1984), and its etiology is likely to be due to a mass displacement effect. Among the more common changes are anterior displacement of the mandible by a massive parotid lesion, outer calvarial changes of the skull, flattening of the nasal pyramid by a large glabellar lesion, and orbital expansion (Figs. 2.28, 2.29).

Cartilaginous destruction also appears to be more common, but no reported incidence has been published. The etiology is unlikely to be simply pressure atrophy and is probably related to enzymatic destruction or ulceration. The two most frequent sites of destruction appear to be the fibrocartilage of the nasal alae and the pinna (Figs. 2.30, 2.31). Conversely, a massive parotid hemangioma may result in overgrowth of the pinna presumably as a consequence of an increased blood supply.

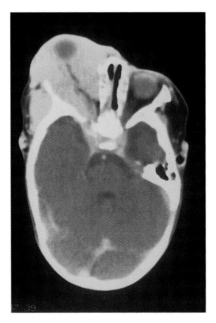

Figure 2.28
A computed tomographic scan showing extensive orbital involvement with proptosis and distraction of the bony margin orbit. The sheer mass of the hemangioma was responsible for this displacement.

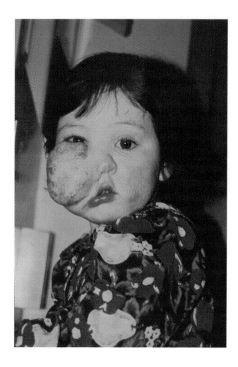

Figure 2.29
A large hemangioma involving the child's right cheek. This lesion displaced her zygomatic arch.

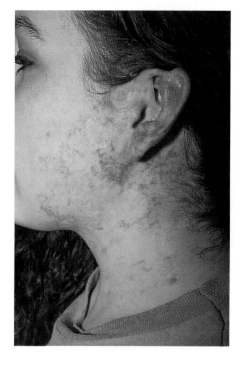

Figure 2.30
An 18-year-old girl with partial loss of her helix.

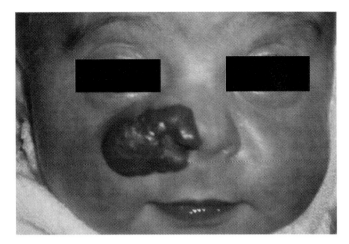

Figure 2.31
An extensive hemangioma of the upper lip. Her paranasal cartilage had ulcerated and destroyed the alum.

2.4. SYSTEMIC HEMANGIOMATOSIS

The presence of multiple cutaneous hemangiomas should alert the clinician to the potential of visceral involvement, in particular hepatic involvement (Fig. 2.32). This condition, known as *systemic hemgiomatosis,*

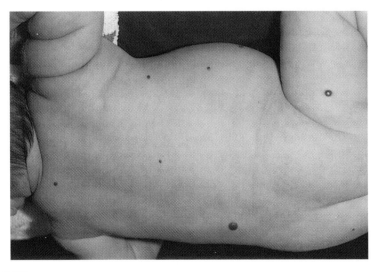

Figure 2.32
An infant with systemic hemangiomatosis. Note the presence of more than 6 cutaneous hemangiomas.

is important to diagnose early because aggressive intervention is necessary and can be life saving. An abdominal ultrasound or magnetic resonance scan is usually diagnostic.

REFERENCES

Amir, J., Metzger, A., Krikler, R., and Reisner, S.H.: Strawberry hemangioma in preterm infants. Pediatr. Dermatol. 3:331–332, 1986.

Azizkhan, R.G., Azizkhan, J.C., Zetter, B.R., Folkman J.: Mast cell heparin stimulates migration of capillary endothelial cells in vitro. J. Exp. Med. 152:931, 1980.

Boon, L.M., Enjolras, O., and Mulliken, J.B.: Congenital hemangiomas: Evidence for accelerated involution. J. Pediatr. 128:329–335, 1996.

Bowers, R.E., Graham, E.A., and Tomlinson, K.M.: The natural history of the strawberry nevus. Arch. Dermatol. 82:667, 1960.

Boyd, J.B., Mulliken, J.B., Kaban, L.B., Upton, J., and Murray, J.E.: Skeletal changes associated with vascular malformations. Plast. Reconstr. Surg. 74:789, 1984.

Brodsky, L., Yoshpe, N., and Rulaen, R.: Clinical–pathological correlates of congenital subglottic hemangiomas. Ann. Otol. Rhinol. Laryngol. (suppl.) 92:108, 1983.

Cheung, D.S.M., Warman, M.L., and Mulliken, J.B.: Hemangioma in twins. Ann. Plast. Surg. 38:269–274, 1997.

Cooper, A., and Bolande, R.: Multiple hemangiomas in an infant with cardiac hypertrophy. Pediatrics 35:27–33, 1965.

Cooper, M., Slovis, T., Madgy, D., and Levitsky, D.: Congenital subglottic hemangioma: Frequency of symmetric subglottic narrowing on frontal radiographs of the neck. AJR 159:1269, 1992.

David, T., Evans, D., and Stevens, F.: Haemangioma with thrombocytopenia (Kasabach-Merritt syndrome). Arch. Dis. Child. 58:1022, 1983.

Deans, R., Harris, G., and Kivlin, J.: Surgical dissection of capillary hemangiomas. Arch. Ophthalmol. 110:1743, 1992.

Dunlap, E.A.: Current effects of ambylopia. Am. Orthop. J. 21:5, 1971.

Dupont, C., Chabrolle, J., De Montis, G., et al.: Angiome cutane et insuffisance cardiaque. Ann. Pediatr. (Paris). 24:37–42, 1997.

Enjolras, O., Wassel, M., Mazoyer, E., Frieden, I., Rieu, P., Drouet, L., Taieb, A., Stalder, J., and Escande, J.: Infants with Kasabach-Merritt syndrome do not have "true" hemangiomas. J. Pediatr. 130:631, 1997.

Esterly, N.: Kasabach-Merritt syndrome in infants. J. Am. Acad. Dermatol. 8:504, 1983.

Finn, M.C., Glowacki, J., and Mulliken, J.B.: Congenital vascular lesions: Clinical application of a new classification. J. Pediatr. Surg. 18:894, 1983.

Healy, G.B., Fearon, B., French, R., and McGill, T.: Treatment of subglottic hemangioma with the carbon dioxide laser. Laryngoscopy 90:809, 1980.

Helikson, M., Shapiro, D., Seashore, J.: Hepatoportal arteriovenous fistula and portal hypertension in an infant. Pediatrics 60:920, 1977.

Hidano, A., and Nakajima, S.: Earliest features of the strawberry mark in the newborn. Br. J. Dermatol. 87:138, 1972.

Hollinger, L., Toriumi, D., and Anandappa, E.: Subglottic cysts and asymmetrical subglottic narrowing on neck radiograph. Pediatr. Radiol. 18:306, 1988.

Holmdahl, K.: Cutaneous hemangiomas in premature and mature infants. Acta Paediatr. 44:370, 1955.

Jacobs, A.H.: Strawberry hemangiomas: The natural history of the untreated lesion. Calif. Med. 86:8, 1957.

Kasabach, H.H., and Merritt, K.K.: Capillary hemangioma with extensive purpura. Am. J. Dis. Child. 59:1063, 1940.

Kessler, D.A., Langer, R.S., Pless, N.A., Folkman, J.: Mast cells and tumor angiogenesis. Int. J. Cancer 18:703–709, 1976.

Koerper, M., Addiego, J.E., deLorimier, A., Lipow, H., Price, D., and Lubin, B.: Use of aspirin and dipyridamole in children with platelet trapping syndromes. J. Pediatr. 102:311, 1983.

Kontras, S.B., Green, O.C., King, L., and Duran, R.J.: Giant hemangioma with thrombocytopenia. Arch. Dermatol. 82:148, 1960.

Kushner, B.: The treatment of periorbital infantile hemangioma with intralesional corticosteroid. Plast. Reconstr. Surg. 76:1985, 517.

Larcher, V.F., Howard, E., and Mowat, A.: Hepatic hemangioma: Diagnosis and management. Arch. Dis. Child. 6:7, 1981.

Lister, W.A.: The natural history of strawberry naevi. Lancet 1:1429, 1938.

Margileth, A.M., and Museles, M.: Cutaneous hemangiomas in children: Diagnosis and conservative management. JAMA 194:135, 1965.

Mulliken, J.B.: Vascular malformations of the head and neck. Mulliken, J.B., and Young, A.E. (eds.): *Vascular Birthmarks: Hemangiomas and Vascular Malformations.* Philadelphia: W.B. Saunders, 1988.

Mulliken, J.B., and Glowacki, J.: Hemangiomas and vascular malformations in infants and children: A classification based on endothelial characteristics. Plast. Reconstr. Surg. 69:412, 1982.

Ogle, J., Hope, R., Watson, C.: Kasabach-Merritt syndrome with terminal gram negative infection. N.Z. Med. J. 83:441, 1976.

Pasyk, K.A., Cherry, G.W., Grabb, W.C., Sasaki, G.H. Quantitative evaluation of mast cells in cellularly dynamic and adynamic vascular malformations. Plast. Reconstr. Surg. 73:69–75, 1984.

Payne, M.M., Moyer, F., Marcks, K.M., and Trevaskis, A.E.: The precursor to the hemangioma. Plast. Reconstr. Surg. 38:64, 1966.

Pereyra R., Andrassy R., and Mahour G.: Management of massive hepatic hemangiomas in infants and children: A review of 13 cases. Pediatrics 70:254–258, 1982.

Pratt, A.G.: Birthmarks in infants. Arch. Dermatol. 67:302, 1967.

Robb, R.M.: Refractive errors associated with hemangiomas of the eyelids and orbit in infancy. Am. J. Ophthalmol. 83:52, 1977.

Sademann, H., and Tygstrup, I.: Prolonged obstructive jaundice and haemangiomatosis. Arch. Dis. Child. 49:665, 1974.

Shim, W.K.T.: Hemangiomas of infancy complicated by thrombocytopenia. Am. J. Surg. 116:896, 1968.

Simpson, J.R.: Natural history of cavernous hemangiomata. Lancet 2:1057, 1959.

Stigmar, G., Crawford, J.S., Ward, C.M., and Thomson, H.G.: Ophthalmic sequelae of infantile hemangiomas of the eyelids and orbit. Am. J. Ophthalmol. 85:806, 1978.

Sutherland, D.A., and Clark, H.: Hemangioma associated with thrombocytopenia. Report of a case and review of the literature. Am. J. Med. 33:150, 1962.

Takahashi, K., Muliken, J., Kozakewich, H., Rogers, R., Folkman, J., and Ezekowitz, R.: Cellular markers that distinguish the phases of hemangioma during infancy and childhood. J. Clin. Invest. 93:2357–2364, 1994.

Thomson, H.G., Ward, C.M., Crawford, J.S., and Stigmar, R.: Hemangiomas of the eyelid: Visual complications and prophylactic concepts. Plast. Reconstr. Surg. 63:641, 1979.

Tsang, W.Y., Chan, J.K.: Kaposi-like hemangioendothelioma: A distinctive vascular neoplasm of the retroperineum. Am. J. Surg. Pathol. 15:982–989, 1991.

Tsang, W.Y., Chan, J.K., Fletcher, C.D.: Recently characterized vascular tumours of skin and soft tissue. Histopathology 19:489, 1991.

Vaksman, G., Rey, C., Marache, P., et al.: Severe congestive heart failure in newborns due to giant cutaneous hemangioma. Am. J. Cardiol. 60:392–394, 1987.

Von Noorden, G.K.: Factors involved in the production of amblyopia. Br. J. Ophthalmol. 58:158, 1974.

Waner, M., Waner, A., Gungor, A., North, P., Mihm, M.: Sites of predilection for hemangiomas. Pediatrics (in press) 1998.

Wishnich, M.: Multinodular hemangiomatosis with partial biliary obstruction. J. Pediatr. 92:960, 1978.

Zukerberg, L.R., Nickoloff, B.J., and Weiss, S.W.: Kaposiform hemangioendothelioma of infancy and childhood. An aggressive neoplasm associated with Kasabach-Merritt syndrome and lymphangiomatosis. Am. J. Surg. Pathol. 17:321–328, 1993.

The Natural History
of Vascular Malformations

MILTON WANER, M.D., F.C.S. (S.A.) AND
JAMES Y. SUEN, M.D., F.A.C.S.

V ascular malformations are true developmental abnormalities that were believed to result from a sporadic, nonfamilial developmental error in the formation of vascular tissue (Mulliken and Glowacki, 1982). However, recent evidence has shed more light on this. Vikkula et al. (1996) found an activating mutation at position R894W resulting in increased activity of tyrosine kinase TIE2 in two families with venous malformations. TIE2 is essential for early vascular development, and an increase in TIE2 activity may lead to abnormal sprouting, branching, and modeling of the primary vascular plexus (Dumont et al., 1994; Sato et al., 1995). In addition, several denominators common to vascular malformations are worthy of consideration. At least two types of malformation are probably due to an abnormality in vascular neural modulation (Smoller and Rosen, 1986; Rydy et al., 1991; Orten et al., 1996; Baker et al., 1998).

Hemangiomas and Vascular Malformations of the Head and Neck, Edited by Waner, M.D. and Suen, M.D.
ISBN 0471-17597-8   © 1999 Wiley-Liss, Inc.

Venular malformations (port-wine stains) are probably due to a relative or absolute deficiency in autonomic innervation of the postcapillary venular plexus, and arteriovenous malformations are possibly also due to the same deficiency but at the level of the precapillary sphincters (Baker et al., 1998). Now it seems that venous malformations are due to an uncoupling of proliferating endothelial cells and smooth muscle cells (Vikkula et al., 1996). Could this type of neural deficiency be responsible for venous malformations as well? Another common feature is that all vascular malformations involve collecting system vessels (i.e., capillaries, venules, veins, and lymphatic vessels) and not arteries. While peripheral arterial abnormalities are known to arise from persistant embryonic arteries, a variability in medium-sized vessel anatomy (an anomalous artery) or an aneurismic dilation of a vessel wall due to a connective tissue disorder (Marfan syndrome, Ehlers-Danlos syndrome, pseudoxanthoma elasticum, and arterial fibrodysplasia). Due to their later presentation and/or their association with other, sometimes more pressing clinical problems, these abnormalities are not commonly dealt with in a vascular anomalies clinic. Furthermore, their etiology appears to be unrelated. More severe arterial malformations would result in agenesis of the relevant structure because a normal arterial pattern seems to be essential for normal structural development.

Vascular malformations are always present at birth, even though they may not be obvious and only manifest at a later date (Sako and Varco, 1970; Finn et al., 1983). They are found with equal frequency in males and females and appear to be as common in all racial groups (Finn et al., 1983). Unlike hemangiomas, their rate of endothelial cell turnover is normal (Mulliken and Glowacki, 1982). They therefore neither proliferate nor involute. Instead, a slow, steady increase in size and/or thickness of the lesion will take place due to hyperplasia. With advancing age, progressive ectasia of the vascular component will lead to expansion of the lesion and in some instances to nodularity (Barsky et al., 1980). Clinical experience has shown that the rate of hypertrophy varies (Barsky et al., 1980). In addition, certain phenomena such as trauma, sepsis, hormonal modulation, and changes in blood or lymph pressure may effect a more rapid or even sudden expansion of the lesion. Apparent fluctuations in the size of lymphatic malformations are thought to be due to the recannalization of lymphaticovenous anastomoses, and, contrary to belief, they never regress completely (Grabb et al., 1980). Table 1.4 summarizes the clinical and biological differences between hemangiomas and malformations.

Mulliken and Glowacki (1982) classified vascular malformations according to the type of vessel making up the lesion. They recognized malformations of all of the primary vessel types and classified malformations accordingly. We thus have capillary, venous, arterial, lymphatic, and mixed malformations comprising two or more vessel types (e.g., venous–lymphatic malformations, venous–lymphatic–capillary malformations, and venous–venular malformations) (Mulliken and Glowacki, 1982). Fifteen years later, this biological classification is still largely intact and widely recognized. Ongoing research in the form of a clinicopathological correlation has led to a refinement of the original classification (North et al., unpublished data, 1998) (see Chapter 5). Because port-wine stains, originally thought to be capillary malformations (Mulliken and Glowacki, 1982), are in fact composed of postcapillary venules, the term *venular malformation* is more appropriate. Furthermore, the nidus of an arteriovenous malformation is an ectatic capillary bed, and thus the term *capillary malformation* is more appropriate. However, because this would create an inordinate amount of confusion, in this volume we continue to refer to them as *arteriovenous malformations,* although we realize that this is a misnomer and that it ultimately should be changed. We have therefore dropped the category of "capillary malformations" for now because, apart from arteriovenous malformations, we have not found a true capillary malformation (see Tables 1.1 and 3.1) (Waner and Suen, 1995).

TABLE 3.1

Classification of Vascular Malformations of the Head and Neck

Venular malformations
Midline venular malformations
Venular malformations
Venous malformations
Lymphatic malformations
Macrocystic lymphatic malformations
Microcystic lymphatic malformations
Arteriovenous malformations
(capillary malformations)
Mixed malformations
Mixed venous venular malformations
Mixed venous lymphatic malformations

3.1. VENULAR MALFORMATIONS

Confusing terminology has prevented us from fully appreciating the clinical scope and behavior of venular malformations. The term *nevus flammeus* has been used collectively to describe two distinct entities: midline venular malformations (nevus flammeus neonatrum or nevus simplex) and venular malformations (port-wine stains). Because both are similar in appearance, these entities are often confused and misdiagnosed. Indeed, the biological classification proposed by Mulliken and Glowacki (1982) considered them both to be "capillary" malformations.

3.1.1. Midline Venular Malformations

Midline venular malformations are commonly referred to by the synonyms *salmon patch, stork bite,* or *angel's kiss.* Unlike their counterpart, the much less common but better-known port-wine stain, these lesions are usually transient and often fade within the first year of life. They always involve midline structures, and the nape of the neck is the most common site (Fig. 3.1) (accounting for 40% in white and 30% in black infants). This is followed by anterior midline lesions (involving the upper eyelids, forehead, glabellum, nasal alae, and philtrum of the upper lip) and then the lower lumbar sacral area (Pratt, 1967). Anterior midline lesions have a characteristic distribution (Fig. 3.2). The forehead and glabellum lesions are typically V shaped and correspond with the distribution of the supratrochlear and supraorbital nerves. Involvement of the nose is typically in the supra-alar crease, and the lip involvement is typically in the upper two-thirds of the philtrum.

Midline venular malformations usually present as a light pink macule, which may be confluent with distinct borders, or nonconfluent. Anterior midline lesions are more often nonconfluent, whereas posterior midline lesions are confluent. Although it is commonly held that most of these will fade within the first year of life, Oster and Nielson (1970) were only able to document a 65% rate of disappearance in males and a 53.8% disappearance rate in females with lesions of the nape of the neck. The rate of disappearance in other sites is probably higher, but this has not yet been documented. Unlike their counterpart, the common port-wine stain (venular malformations), midline lesions almost never progress. Hypertrophy and cobblestone formation are thus extremely rare.

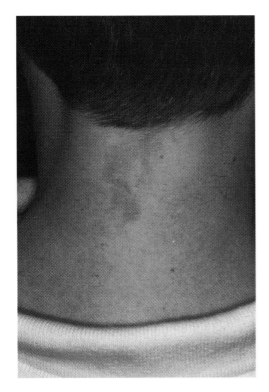

Figure 3.1
A child with a persistent midline venular malformation involving the nape of his neck.

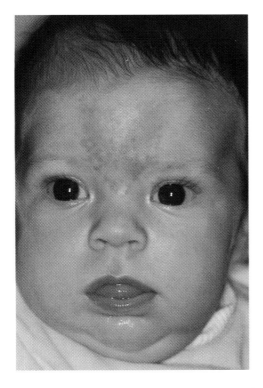

Figure 3.2
An infant with a midline venular malformation involving her forehead and upper lip. The typical V-shaped distribution of this lesion is evident, as well as involvement of the midline philtrum and the alum of the left nostril.

Histologically, Schnyder (1955) was unable to demonstrate vascular ectasia in affected infants, but was able to do so in persistent nuchal lesions in older children. In all probability, midline venular malformations represent an ectasia of postcapillary venules probably on the basis of *delayed* innervation of the plexus by the autonomic nervous system. This probability is supported by the fact that both persistent nuchal stains and anterior midline stains almost never hypertrophy or form cobblestones. In contradistinction to this, port-wine stains represent a progressive ectasia of postcapillary venules due to an absolute or relative deficiency of autonomic innervation to the vessel walls (vs. midline capillary malformations in which there is probably a delayed development of the autonomic fibers). The macroscopic correlates of progressive ectasia are thickening and darkening of the lesion, soft tissue hypertrophy, and cobblestone formation.

3.1.2. Venular Malformations (Port-Wine Stains)

Port-wine stains are erroneously referred to as *capillary hemangiomas.* These lesions are histologically and clinically true malformations and are made up of postcapillary venules within the papillary plexus. The precise etiology of port-wine stains is unknown. Barsky et al. (1980) thought that they probably represent progressive ectasia of a once-normal cutaneous vascular plexus and that this ectasia is progressive with advancing age, hence the darkening in color, hypertrophy, and eventual cobblestone formation. Smoller and Rosen (1986) postulated that an altered or even absent neural modulation of the vascular plexus was responsible for this progressive ectasia. An analysis of biopsy specimens from both normal skin and port-wine-stained skin showed no difference in vessel number. There was, however, a marked decrease in the perivascular nerve density in port-wine-stained biopsy specimens. Rydy et al. (1990) demonstrated a decrease in both sympathetic and sensory innervation of the papillary plexus. Because sensory fibers elaborate several neuropeptides, including substance P, which is known to stimulate smoothe muscle growth and calcitonin gene–related peptide (which stimulates endothelial cell growth), they postulated an absence of trophic effects as a possible cause (Nilsson et al., 1985; Haegerstrand et al., 1990). In light of all these findings, we have postulated the "sick dermatome" theory to explain the etiology of these lesions. We believe that a venular malformation is a manifestation of a "sick dermatome" in which there is an absolute or relative deficiency of vascular autonomic and sensory vascular innervation as the underlying pathology. Lesions with an absolute deficiency will progress more rapidly and early hypertrophy with cobblestone formation is likely, whereas a relative deficiency of autonomic innervation will give rise to slower progression.

Port-wine stains occur in 0.3% of the population, and the male to female ratio is 1:1 (Jacobs and Walton, 1976). They are present at birth, but may not be apparent for at least a few days. The usual presentation is that of a flat, pink macule, and most (83%) occur in the head and neck (Orten et al., 1996). Curiously, in at least one large series, right-sided lesions were twice as common as left-sided lesions (Barsky et al., 1980). Venular malformations partially or completely affect one or more sensory dermatomes, and in the head and neck this means one or more of the trigeminal dermatomes (Figs. 3.3, 3.4). The V2 dermatome is the most commonly involved (57%), followed by the mandibular dermatome (V3) and then the ophthalmic (V1) (Orten et al., 1996).

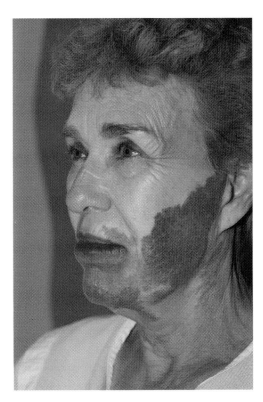

Figure 3.3
A patient with involvement of the V3 dermatome. This patient has a confluent lesion.

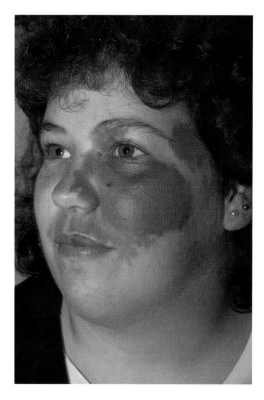

Figure 3.4
A patient with involvement of the
V2 dermatome. This patient has a
confluent lesion.

In lesions where more than one dermatome was involved, V2 was involved with V1 and/or V3 in 90% of cases. Within a dermatome, the lesion may be confluent or stippled (geographical) (Figs. 3.4, 3.5). This is
important because "geographical" lesions are more likely to respond
well to treatment (Orten et al., 1996). Mucosal involvement adjacent to
or in continuity with a cutaneous lesion is common, especially in the
presence of a midfacial V2 lesion, where the vermillio–cutaneous border, lip mucosa, and maxillary gingiva may be involved (Fig. 3.6).

Unlike midline venular malformations, venular malformations persist and darken and thicken with age. Thus, a light pink lesion at birth
may soon darken to a deep red lesion by puberty and progress to a
dark purple, thick, cobblestoned port-wine stain by 30 years of age
(Figs. 3.7, 3.8). Furthermore, the rate of this progression varies. One can
therefore encounter light pink lesions in adults and dark purple lesions
in children. Histologically, Barsky et al. (1980) compared the mean vessel area with age and depth in color and concluded that progression in
color and in thickness is a direct result of progressive vascular ectasia.

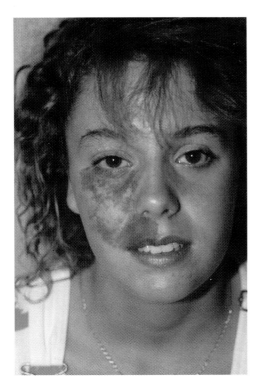

Figure 3.5
A patient with involvement of the V2 dermatome. This patient has a geographical pattern of involvement and is much more likely to respond well to laser treatment.

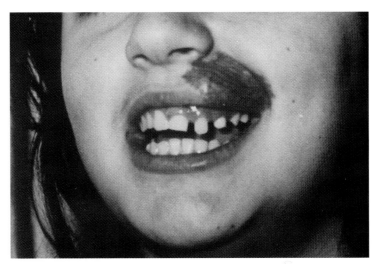

Figure 3.6
A teenager with involvement of the skin of her upper lip, vermilion, mucosa, and gingiva. As a consequence of gingival involvement, she also has hypertrophy of her maxilla. Increased spacing of her ipsilateral teeth has resulted from this.

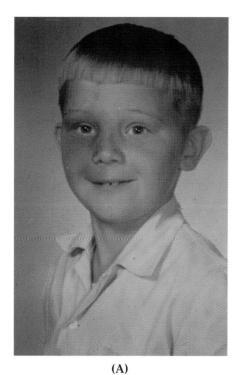

(A)   **(B)**

Figure 3.7
(A) A child with a light pink macule involving the right V2 and V3 dermatomes.
(B) The same child as a teenager with a much darker lesion. This lesion progressed
rapidly to cobblestone formation by the time the patient was in his early 20s.

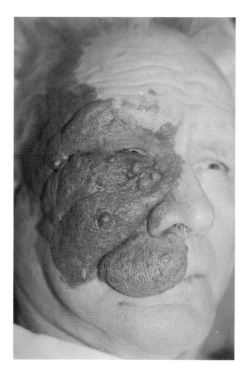

Figure 3.8
A 65-year-old man with a thick, cobblestoned port-wine stain. Note the
obvious hypertrophy of his upper lip
as well as the soft tissues of his upper
and lower eyelid and cheek. This lesion was once a light pink macule.

Variation in the rate of ectasia can be accounted for by the "sick dermatome" theory. According to this theory, there is a relative or an absolute
deficiency in the innervation of the venular plexus; therefore patients in
whom there is an absolute deficiency will experience early and more rapid
progression of their lesions. Early darkening, thickening, and cobblestone
formation can be expected. Patients in whom there is a relative deficiency
will, on the other hand, experience slow progression of their lesion.

Within the papillary vascular plexus, videomicroscopy has revealed
three patterns of vascular ectasia (Motly et al., 1996) (see Fig. 3.9): type
1, ectasia of the vertical loops of the papillary plexus; type 2, ectasia of
the deeper, horizontal vessels in the papillary plexus; and type 3, mixed
pattern with varying degrees of vertical and horizontal vascular ectasia. Recognition of these patterns is important in that they impact on
the response to treatment. Given that laser light has a limited depth of
penetration, type 1 lesions are more apt to respond well to treatment,
and type 3 lesions are likely to respond poorly.

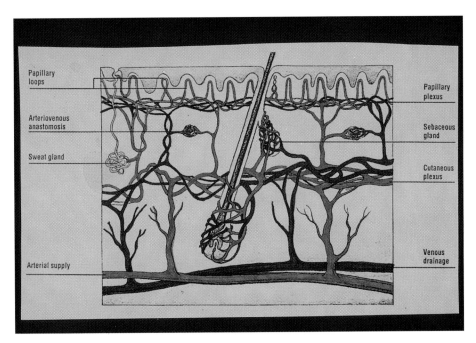

Figure 3.9
A schematic representation of the blood supply to skin. Port-wine stains involve
either the vertical or horizontal vessels of the papillary plexus. Motly et al., 1996
described three patterns of vascular ectasia. Type 1 involves the vertical loops;
type 2, the horizontal vessels; and type 3, a mixed pattern with varying degrees of
both vertical and horizontal vessels. Deep port-wine stains involve the cutaneous
plexus; venous malformations involve the subcutaneous plexus.

Because there is so much variability in the appearance of venular malformations, it became essential to classify these lesions. Earlier attempts at classification relied on color and used paint color charts. These were found to be unreliable because it was difficult to match flesh tones with the colors used to paint the interior of a building. Therefore, this was not an appropriate way of classifying venular malformations. Furthermore, the color of a lesion is determined by the degree of oxygenation of the hemoglobin, which is dependent on the degree of perfusion of the vascular bed, which in turn depends on a number of factors such as ambient temperature, the level of circulating catecholamines and local metabolites. We now classify venular malformations in accordance with their degree of vascular ectasia because this is the true cause of the variation in clinical appearance. This classification recognizes four grades of ectasia, grades 1 to IV. Grade I represented the smallest vessels and grade IV the largest. When using this classification, one should always bear in mind that there is a progression between the grades and that the divisions between grades are, to a large extent, arbitrary. The main purpose of this classification is to assign a grade for ease in communication and determination of treatment modality.

Grade I lesions are the earliest lesions and thus have the smallest vessels (50–80 μm in diameter). Using $\times 6$ magnification and transillumination, individual vessels can only just be discerned and appear like grains of sand. Clinically, these lesions are light or dark pink macules (Figs. 3.7A and 11.2).

Grade II lesions are more advanced (vessel diameter = 80–120 μm). Individual vessels are clearly visible to the naked eye, especially in less dense areas. They are thus clearly distinguishable macules (Fig. 11.3).

Grade III lesions are more ectatic (120–150 μm). The large end on the vessels are visible and impart a reddish color to the lesion. These lesions are still macular (Fig. 11.4).

Grade IV lesions are the most advanced (> 150 μm). By this stage, the space between the vessels has been replaced by the dilated vessels. Individual vessels may still be visible on the edges of the lesion or in a less dense lesion, but by and large individual vessels are no longer visible. The lesion is usually thick, purple, and palpable (Fig. 3.5, 3.6). Eventually, dilated vessels will coalesce to form nodules, otherwise known as *cobblestones* (Fig. 3.8).

Skeletal changes in the form of bony hypertrophy are sometimes seen (Figs. 3.6, 3.10). This is especially true of V2 lesions that extend onto mucosal and gingival surfaces of the maxilla (Fig. 3.6). Localized hypertrophy of the underlying bone, increased spacing between the teeth, and a prominence of the upper lip, due in part to hypertrophy of the underlying alveolar ridge, are thus seen. While soft tissue hypertrophy is common, especially in older patients, the role of bony hypertrophy is often unrecognized, and can lead to unnecessary and ineffective surgical debulking of the upper lip.

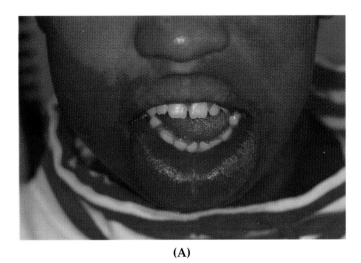

(A)

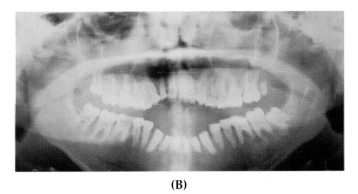

(B)

Figure **3.10**
(A) A child with a diffuse port-wine stain involving the lower third of her face. Note the soft tissue hypertrophy and her asymmetrical open bite. **(B)** An orthopentanogram showing bony hypertropy of the anterior one-third of the ramus of her mandible as well as of her premaxillary area.

3.1.3. Meningeal and Choroidal Involvement

Venular malformations can involve other neurectodermal structures, such as the choroid and meninges. This was first described by Sturge in 1879 and confirmed 18 years later at a postmortem by Kalischer (1897). A venular malformation involving the dermal vascular plexus, with choroidal and ipsilateral meningeal involvement, has become known as the *Sturge-Webber syndrome* (Fig. 3.11). By definition, all three components must be present for the diagnosis. Dermal and choroid involvement without meningeal involvement may also be found but is therefore not Sturge-Weber syndrome. Patients with Sturge-Weber syndrome commonly present with glaucoma and sequelae of neural involvement such as seizures, cerebral atrophy, degeneration, and subdural hemorrhage. However, there is a great variation in the occurrence

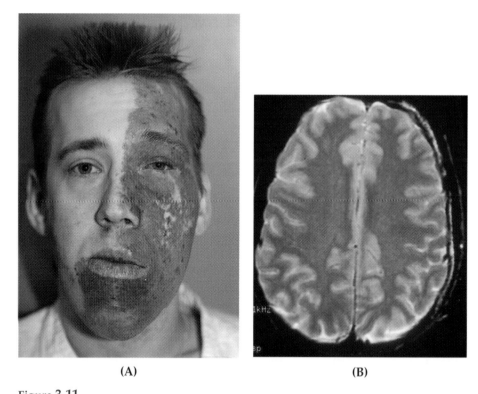

(A)   (B)

Figure **3.11**
(A) A young adult with Sturge-Weber syndrome. This patient also has soft tissue hypertrophy mandibular hyperplasia. **(B)** His magnetic resonance imaging scan shows ipsilateral meningeal involvement.

and the severity of these sequelae. The "sick dermatome" theory can account for this variation. Patients who have an absolute deficiency of venular innervation are more likely also to have severe glaucoma and neurological manifestations, whereas patients who have a relative deficiency are less likely to have these manifestations. The cutaneous component of Sturge-Weber syndrome is histologically identical to that seen in conventional venular malformations except that it is more extensive and frequently involves both sides of the face (Barsky et al., 1980). Furthermore, in almost all cases the lesion is confluent. Hypertrophy of the subjacent maxilla and mandible and soft tissue hypertrophy of the skin and appendages are also more common.

Glaucoma in the absence of intracranial involvement is not an infrequent finding in facial lesions and occurs in approximately 10% of patients (Barsky et al., 1980). The likelihood of glaucoma developing in a patient with V1 and V2 involvement is, however, much greater. Barsky et al. (1980) concluded that the incidence in these cases was 27%, whereas Stevenson et al. (1974) found a much higher rate of occurrence (45%). Barsky et al. (1980) also found that every patient with glaucoma had both V1 and V2 involvement. Every child with eyelid involvement, with or without V1 or V2 involvement, should therefore undergo tonometry at 6-month intervals until age 3 years and yearly thereafter. The etiology of glaucoma in these cases is probably related to an increased ante-episcleral venous pressure that raises the intraocular pressure (Phelps, 1978). Glaucoma in patients with Sturge-Weber syndrome may, on the other hand, be due to a degeneration of the trabecula meshwork of the canal system of Schlem (Cibis et al., 1984).

While the vast majority of venular malformations are isolated, uncomplicated dermal vascular malformations, one should always be aware of the possibility of an underlying associated vascular lesion. In these cases, additional signs will be present and should be excluded. A venous malformation, for example, may be associated with an overlying venular malformation in a mixed venous–venular malformation and will be evident as a soft, compressible mass underlying the venular malformation. In this instance, ectasia of both the subcutaneous venous plexus and the papillary venular plexus is present. By the same token, an underlying arteriovenous malformation may do likewise (Fig. 3.12). A stain on the overlying skin probably means involvement of the overlying skin by the arteriovenous malformation because failure to remove this skin together with the underlying malformation will almost

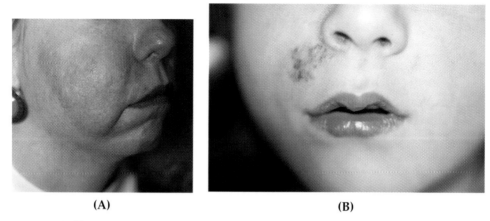

(A)   (B)

Figure **3.12**
(A) An adult with a "vascular blush" overlying an area of fullness involving her right cheek. The lesion turned out to be a low-grade arteriovenous malformation. **(B)** A "vascular blush" overlying a venous malformation of the nasal labial region. The blush is merely a superficial extension of the underlying venous malformation.

invariably result in a recurrence. In these cases, additional clinical signs such as a mass, increased skin temperature, or a palpable thrill will be evident.

3.2. VENOUS MALFORMATIONS

Venous malformations are made up of ectatic veins. Whether these lesions are simply an ectasia of an otherwise normal venous bed or are true developmental anomalies with an increase in vessel number as well is not known. Like all other malformations, venous malformations are progressive. The degree of ectasia increases with advancing age, but the rate at which this takes place is variable (Fig. 3.13). Some lesions progress rapidly, and there is a noticeable increase in size from year to year (Fig. 3.17), whereas in others the changes are slow and several years may lapse before there is a noticeable increase in size. We thus have high-grade and low-grade lesions.

As with any other type of malformation, venous malformations may be superficial or deep, localized, multicentric, or diffuse (Figs. 3.14–3.16, 3.18, 3.19). The overlying skin or mucosa will vary in color,

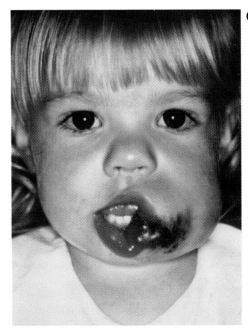

(A)

Figure 3.13
(A) A 3-year-old child with an obvious venous malformation involving her buccal fat space and oral commissure. **(B)** The same child as an infant. Note the considerable growth that this lesion has undergone in the short space of 18 months. This is therefore a high-grade venous malformation.

(B)

depending on the depth and degree of ectasia of the underlying malformation. Because expansion invariably progresses along planes of least resistance, superficial malformations will expand toward the sur-

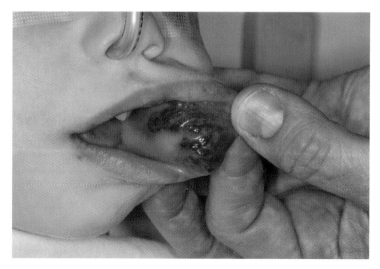

Figure 3.14
A localized venous malformation of the buccal mucosa.

Figure 3.15
A teenager with a venous malformation involving her buccal fat space. Because the lesion is deep, no surface discoloration is visible.

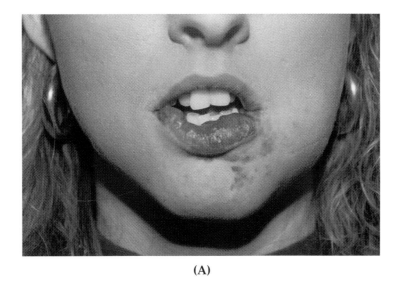

(A)

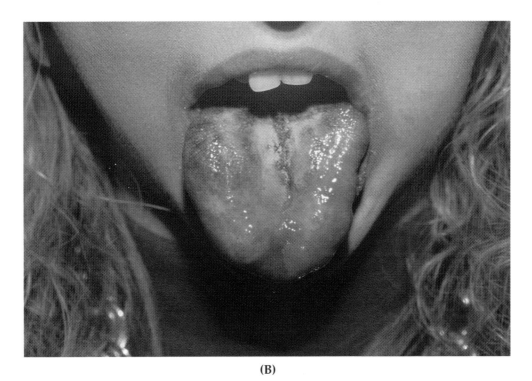

(B)

Figure **3.16**
(A,B) A young adult with a multicentric venous malformation involving her tongue, lower lip, and the skin of her chin.

Figure 3.17
A child with a massive venous malformation of the tongue and cervicofacial region. This has resulted in considerable soft tissue hypertrophy of the tongue.

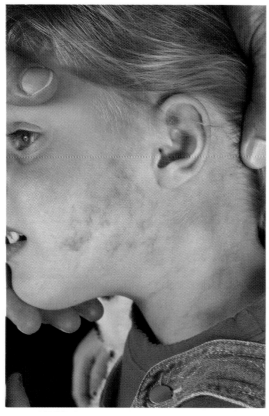

Figure 3.18
A diffuse cervicofacial venous malformation. Although minimal surface discoloration is evident, this diffuse lesion involves the lower third of her face and almost two thirds of her neck.

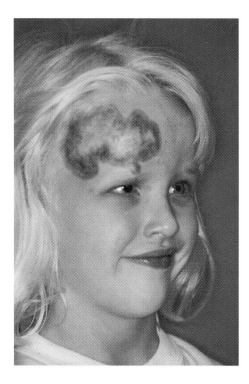

Figure **3.19**
A child with an arteriovenous malformation involving the right side of her forehead. Note that, as a consequence of a steal syndrome, the skin overlying the lesion had ulcerated. She eventually healed but remained with atrophic scarring.

face of the skin or the mucosal surface and impart a deep purple discoloration (Fig. 3.14). Furthermore, nodularity, commonly seen with portwine stains, is also a feature of superficial venous malformations. Deeper malformations impart varying degrees of a bluish hue—the deeper the lesion, the less the discoloration (Fig. 3.15). Very deep lesions such as those of the buccal fat space appear as a flesh-colored mass. The overlying skin is therefore not discolored. Venous malformations are usually soft and compressible. Increased venous pressure, seen during crying, straining, or maintaining the lesion in a dependent position, will expand the lesion. Because flow is inversely proportional to the square of the radius (f α $1/r^2$), progressive ectasia will result in a considerable decrease in the flow rate. Phlebothrombosis thus occurs with greater frequency in ectatic lesions, and phleboliths are considered a hallmark of venous malformations. The thrombis may become infected and cause rapid expansion of the lesion with pain and tenderness not dissimilar to that with thrombophlebitis. Rapid expansion may also follow partial surgical resection, trauma, and hormonal modulation (Mulliken and Young, 1988).

Venous malformations may be unifocal or multifocal, and certain sites within the head and neck are more commonly involved. These include the buccal mucosa, the tongue, the oral commissure, the lower or upper lip, the buccal fat space, and the neck (Figs. 3.13–3.16). Within these anatomical sites, venous malformations do not appear to respect anatomical borders. Muscle, salivary gland, and skin infiltration are common. Venous malformations may, on occasion, be entirely intraosseus. Although the mandible is the most common site, maxillary, nasal parietal, and frontal malformations have been reported (Kirchoff et al., 1978; 1985; Schmidt, 1982). Sadowsky et al. documented mandibular and maxillary involvement and referred to these as *central hemangiomas* (see Schmidt, 1982). Mandibular lesions usually appear as painless, slow-growing masses. Increased spacing between the teeth and increased mobility of the teeth due to expansion of the buccal cortex and/or bleeding around the gingival necks of the teeth are all manifestations (Schmidt, 1982). The first sign of an intraosses venous malformation may, however, be profuse, unexpected hemorrhage following a tooth extraction. Radiological confirmation by way of a soap bubble or honeycomb appearance within the bone will usually alert one to the diagnosis. Apart from frank intraosses involvement, bony hypertrophy and/or distortion are not uncommon sequelae (Barsky et al., 1980; Boyd et al., 1984). Several mechanisms may be operative, and these may be mechanical, physiological, and developmental. Mechanical factors are probably responsible for bony distortion, but hypertrophy may be explained on the basis of an altered blood flow that in turn leads to an elevated temperature and oxygen concentration. Soft tissue hypertrophy is also not uncommon, especially with more diffuse lesions (Fig. 3.17).

Vikkula et al. (1996) demonstrated an activating mutation of position R894W that in turn led to an increase in the receptor tyrosine kinase TIE2 activity in two families with congenital venous malformations. This, together with their immunohistochemical analyses, which showed that the vessels (which were up to 10 times the diameter of normal vessels) were surrounded by an extremely variable thickness of smooth muscle, suggests abnormal vessel development caused by a local uncoupling between proliferating and differentiating endothelial cells and smooth muscle cells (Vikkula et al., 1996). Whether this mechanism accounts for all venous malformations remains to be seen.

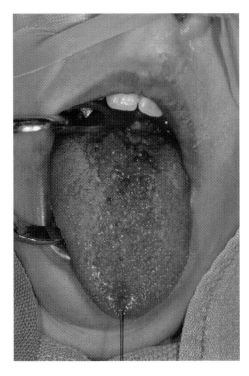

Figure 3.20
Lymph-filled vesicles involving the anterior two-thirds of the tongue of this patient. These are manifestations of a superficial lymphatic malformation or the superficial component of a deep malformation.

3.3. LYMPHATIC MALFORMATIONS

Congenital defects of the lymphatic system constitute a spectrum of disorders that manifest in a variety of ways. These include lymphedema, chylous effusions, localized gigantism, and soft tissue tumors. All congenital malformations of the lymphatic system that result in a contiguous mass of dilated lymphatics are known as a *lymphatic malformations*. This term has replaced previous ones such as *lymphangioma, cystic hygroma, lymphangioma circumscriptum,* and *lymphangiomatosis.* Primary lymphedema results when there is a deficiency or a defect in the lymphatic channels, and secondary lymphedema occurs when the lymphatic channels are normal but the regional lymph nodes are affected. Depending on the severity of the defect, lymphedema can manifest early or late. In milder defects, the lymphatic system may cope well under normal circumstances and only fail during stressful events such as infection, trauma, and pubertal hormonal surges, which result in an increased volume of lymph. In lymphatic malformations, the primary defect is believed to be at the level of the efferent channels. Obstruction at this level, whether relative or absolute, will result in dilation of the proximal channels that form the mass (Levin, 1989).

Although all lesions, by definition, are present at birth, some only become manifest at a later stage. From 65% to 75% are diagnosed at birth, and 80%–90% are diagnosed by the end of the second year of life (Gross, 1953). The remainder may manifest as late as the first pregnancy or even later. In the presence of a mild obstruction (a low-grade lesion), the lymphatic system will clear a normal fluid load, but will fail when faced with some form of stress that results in an increased volume of lymph. This can take the form of an infection, trauma, or hormonal surge. Alternatively, radiation therapy or some connective tissue disorder can place a burden on the system by exacerbating the obstruction or increasing the volume of lymph—hence the late presentation.

The head and neck region is the most common site for lymphatic malformations, and over 90% are found on the neck (Ward et al., 1950). The complexity of the cervical lymphatic system has been offered as a possible explanation (Kennedy, 1989). Clinical manifestations vary in accordance with the extent and depth of the lesion as well as the degree of fibrous reaction around it. Mucosal or cutaneous involvement usually results in the formation of multiple cutaneous or mucosal fluid-filled vesicles (Figs. 3.20, 3.21). These vesicles may be connected to larger, deeper lymphatic cisterns lying within the subcutaneous or the submucosal tissues. The involvement of deeper tissues follows one of two patterns, either massive generalized edema with poorly defined borders (the diffuse *microcystic variety*) or a localized area of multilocular cysts (the *macrocystic* variety). Generally, cervical lesions (previously known as *cystic hygromas*) tend to be localized, macrocystic, and thus more amenable to complete surgical resection. Malformations involving the floor of the mouth, the cheeks, and the tongue are more inclined to be poorly defined, diffuse, microcystic, edematous lesions that are rarely amenable to complete resection (Figs. 3.22, 3.23).

As with other vascular malformations, the rate of growth varies. In general, the earlier the diagnosis, the more aggressive the lesion. These are high-grade lesions. Low-grade lesions present later and are less inclined to result in complications. Sudden or rapid expansion may result from sepsis and spontaneous or traumatic interlesional hemorrhage (Broomhead, 1964). Both of these complications are frequent, and their

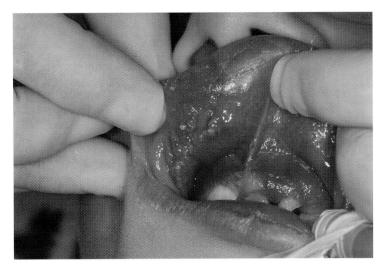

Figure 3.21
The buccal mucosa of a child demonstrating the presence lymph-filled vesicles. This is a manifestation of a superficial lymphatic malformation or the superficial component of a deep malformation.

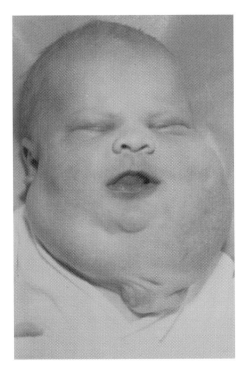

Figure 3.22
An infant with an extensive cervicofacial lymphatic malformation. The lesion resulted in significant airway obstruction and this necessitated a tracheostomy.

abrupt onset may precipitate a life-threatening emergency due to air-way obstruction. Ninh reported a 16% incidence of sepsis (Ninh and Ninh, 1974). Mulliken and Young (1988) postulated the inability of anomalous lymphatic tissue to handle the frequent seeding of oral bacteria as a possible cause for the high rate of sepsis. The incidence of spontaneous hemorrhage is reported to be in the region of 8% in cystic lesions (Broomhead, 1964). Erosion into an intralesional blood vessel by an expanding lesion may well explain this phenomenon (Broomhead, 1964). Respiratory distress is also a frequent complication and may present either as a slow insidious progression of stridor or as a sudden profound obstruction due to sepsis or intralesional hemorrhage. Lesions of the floor of the mouth may displace the oral contents posterosuperiorly, resulting in both airway and pharyngeal obstruction. Cervical lesions, on the other hand, may displace and eventually compress the pharynx, and mediastinal extension may compress the trachea (Fig. 3.22) (Kennedy, 1989). Endolaryngeal cystic mucosal blebs may also embarrass and on occasion, totally obstruct the airway. Both

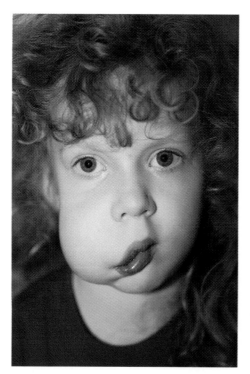

Figure 3.23
A 2.5-year-old child with a large lymphatic malformation involving her buccal fat space and upper lip. The lesion has caused considerable facial distortion, with lengthening of her upper lip.

soft tissue and skeletal hypertrophy are extremely common. Macroglossia, macrotia, and macrohelia are all well-documented sequelae of lymphatic malformations (Fig. 3.24). Boyd et al. (1984) noted that skeletal hypertrophy occurred in 83% of cases and skeletal distortion in 33% of cases. Since lymphatic malformations are not associated with an increased blood supply, skeletal hypertrophy is not likely to be a vascular phenomenon. Mandibular hypertrophy will give rise to prognathism and malocclusion (Fig. 3.24). (Knowles, 1971). Histological features include dilated lymphatic channels or cysts, lined with a single layer of flat endothelium, in a bed of dense fibrous connective tissue with follicles of lymphocytes and occasional germinal centers. Macrocystic lesions are made up of large cysts lined by one or more layers of endothelium and tend to be localized. Macrocystic, diffuse lesions were thought of as being infiltrative because they were connected with finger-like projections into adjacent structures. This underscores the diffuse, poorly localized nature of this variety of lymphatic malformation.

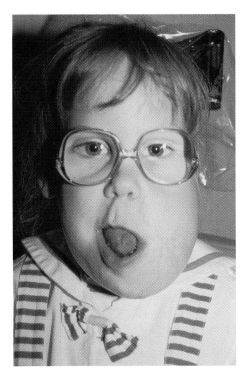

Figure 3.24
An 8-year-old child with macroglossia and mandibular hypertrophy due to a large lymphatic malformation. She has an open bite with tongue protrusion.

Scattered reports of spontaneous regression have prompted several investigators to delay treatment (Kennedy, 1989; Broomhead, 1964; Mulliken and Young, 1988). The only likely explanation for this curious but rare phenomenon is the establishment of venous lymphatic shunts (Sabin, 1909). An incorrect diagnosis may also account for at least some of these cases because considerable difficulty is sometimes encountered in distinguishing between a subcutaneous hemangioma and a lymphatic malformation. Because hemangiomas always involute, this may be mistakenly diagnosed as regression in a lymphatic malformation. An alternate explanation may be that these were low-grade lesions that only manifested during a stressful event associated with an excessive production of lymph.

3.4. ARTERIOVENOUS MALFORMATIONS

The term *arteriovenous malformation* refers to a congenital malformation made up of multiple fistulous tracts communicating between arteries and veins. Mulliken and Young (1988) prefer to reserve the term *arteriovenous fistula* for the acquired, traumatic variety, usually made up of a single fistulous tract. An understanding of the pathiophysiology and the natural history of an arteriovenous malformation is essential to planning their management. Without doubt, the single most important contribution to our understanding of these lesions is that of Holman (1968). A proximal arteriovenous malformation with a significant shunt may cause an increase in cardiac output that in turn may lead to high-output cardiac failure. Distal shunting, on the other hand, has the propensity to significantly reduce the flow rate beyond the shunt and so induce peripheral ischemia (steal syndrome) without adversely affecting the cardiac output. Because an arteriovenous malformation constitutes a high to low pressure shunt, an increased flow in the afferent artery may be seen due to the decreased peripheral resistance. This increased flow will in turn cause dilation and tortuousity of the afferent artery with subsequent thickening of the wall due to hypertrophy in the media (Fig. 3.25). In the face of a large shunt across an arteriovenous malformation, the arterial flow distal to the shunt may reverse direction, thus causing distal ischemia, the so-called steal syndrome (Fig. 3.19). The decrease in arterial pressure also encourages the development of an extensive collateral circulation. An increased flow through

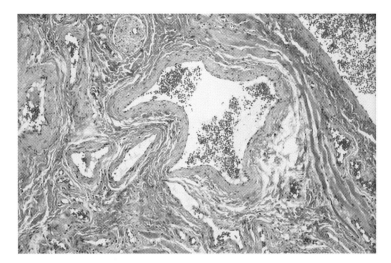

Figure 3.25
An H&E-stained section of an arteriovenous malformation showing large hypertrophied arteries and dilated veins.

the arteriovenous malformation may also dilate the efferent venous
system with consequent thickening of the wall due to hypertrophy of
the media.

Histological analysis of arteriovenous malformations has shed some
light on their pathogenesis (Baker et al., 1998). Examination of specimens from younger children and infants revealed that a nidus is made
up of a bed of dilated capillaries (Fig. 3.26). As the lesion matures, the
degree of ectasia increases, and the development of venous dilation and
arterial hypertrophy becomes apparent (Fig. 3.25). The primary abnormality or nidus therefore appears to be an ectatic capillary bed. Arterial
hypertrophy and venous dilation are merely secondary phenomena that
result from the increased flow across the nidus. Because the nidus is
simply an ectatic capillary bed and because the precapillary sphincters
regulate the blood flow through the capillary bed, we believe that arteriovenous malformations result from an abnormality at the level of the
precapillary sphincter. An absence of autonomic nerve supply to the
sphincters, an absence of the actual sphincters, or some deficiency in the
neuroreceptors at this level will result in free flow across that particular
capillary bed. In time, the vessels in the bed dilate, and eventually the
area supplying the arteries hypertrophy and the veins dilate. This absence of capillary sphincter control may be absolute or relative, hence
the variation in age of presentation and speed of progression.

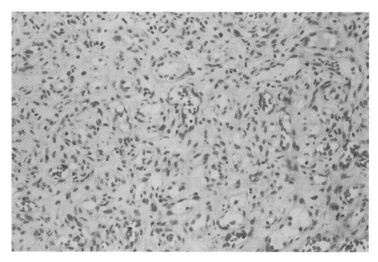

Figure 3.26
An H&E-stained section section of an "early" arteriovenous malformation. At this stage, the lesion is made up of a bed of ectatic capillaries.

Ligation of a feeding artery is therefore ineffective and, indeed, lowers the pressure in the afferent artery distal to the point of ligation. This, in turn, will cause "stealing" due to a reversal of the direction of flow distal to the shunt, thus resulting in distal ischemia. By the same token, ligation of an artery above and below the arteriovenous malformation may well prevent stealing, but will be equally ineffective due to the extent of collateral circulation.

Although all congenital arteriovenous malformations are, by definition, present at birth, clinical presentation is usually delayed. Arteriovenous malformations may thus present during the third, fourth, or even fifth decade of life (Figs. 3.27–3.29) (Sako and Varco, 1970; Mulliken and Young, 1988). A mass complex comprised of dilated tortuous arteries and veins is the most common manifestation of an arteriovenous malformation. Occasionally, an overlying port-wine stain may mask the true diagnosis (Figs. 3.12A, 3.30). Ulceration of the overlying skin and skeletal hypertrophy are all clinical signs (Figs. 3.31., 3.32).

Arteriovenous malformations are firmer to palpation than venous malformations and do not empty quite as readily as venous malformations do with compression. Once compressed, they rapidly refill. One may feel a pulsation or even a fluid thrill, but this is uncommon.

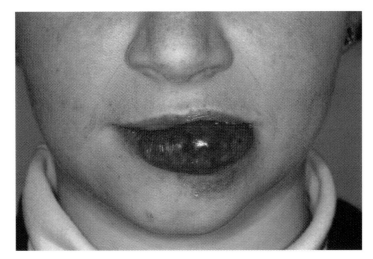

Figure 3.27
A 30-year-old woman with an arteriovenous malformation of her lower lip that had presented for the first time in the third decade of life. The patient claimed that she had bitten her lip during the delivery of her first child and that the lesion had first appeared at that time and progressed from that point onward. Closer questioning, however, elicited the fact that a vascular "blush" had presented for the first time during her teenage years. This lesion turned out to be an arteriovenous malformation.

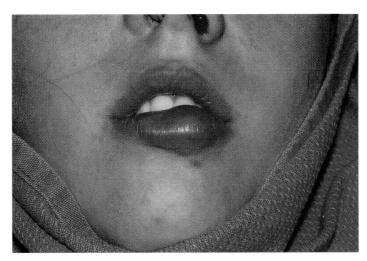

Figure 3.28
An arteriovenous malformation of the lower lip. Although almost identical to the preceding lesion (Fig. 3.27), this child's lesion presented earlier, within the first few years of life. It is therefore a much higher grade lesion.

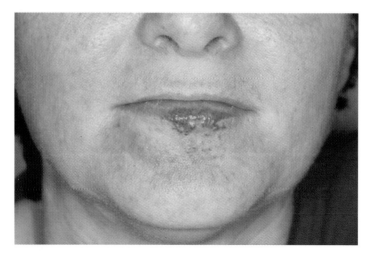

Figure 3.29
A low-grade arteriovenous malformation of the lower lip. This lesion presented for
the first time in the fourth decade.

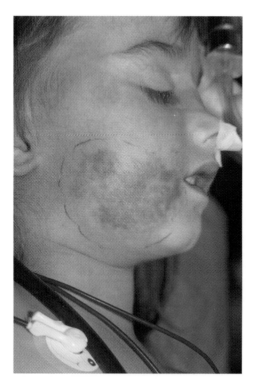

Figure 3.30
An overlying port-wine stain may
mask the presence of an underlying
vascular lesion. In this instance, the
child had an extensive arteriovenous
malformation underlying the cutaneous manifestation. Surgical removal of the underlying lesion only
resulted in a recurrence because skin
was obviously involved.

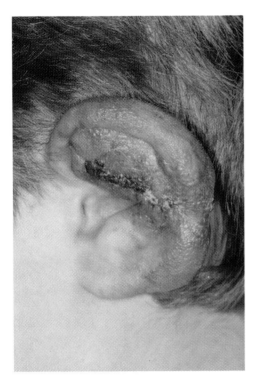

Figure **3.31**
Ulceration of the overlying skin is a common feature of a superficial arteriovenous malformation. In addition, the malformation has resulted in hypertrophy of the soft tissue of the pinna.

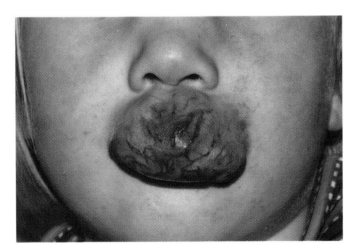

Figure **3.32**
A young child with an arteriovenous malformation over her upper lip. Note the distention and soft tissue hypertrophy. The midline of the upper lip appears to be a site of predilection for arteriovenous malformations.

Most arteriovenous malformations are firm vascular lesions surrounded by hypertrophied arteries and dilated veins. With advancing age, these become much more prominent.

An increased circulating blood volume will counter the effect of decreased arterial pressure and in turn lead to an increased stroke volume, tachycardia, and cardiomegaly. High-output cardiac failure is rare, but has been reported in infants with a large arteriovenous malformation (Flye et al., 1983). Indeed, an otherwise healthy infant can quite adequately compensate for an increased cardiac output for many years before decompensating (Natali et al., 1984).

REFERENCES

Baker, L., Waner, M., Thomas, R., and Suen, JY.: The management of arteriovenous malformations. Arch. Otolaryngol. (in press) 1998.

Barsky, S.H., Rosen, S., Geer, D.E., and Noe, J.: The nature and evolution of port wine stains: A computer assisted study. J. Invest. Dermatol. 74:154, 1980.

Boyd, J.B., Mulliken, J. B., Kaban, L.B., Upton, J., and Murray, J.E.: Skeletal changes associated with vascular malformations. Plast. Reconstr. Surg. 76:789, 1984.

Broomhead, I.W.: Cystic hygroma of the neck. Br. J. Plast. Surg. 17:225, 1964.

Cibis, G.W., Triphathi, R.C., and Tripathi, B.J.: Glaucoma in Sturge-Weber syndrome. Ophthalmology 91:1061, 1984.

Dumont, D., Gradwohl, G., Fong, G., Pur, M., Gertserstein, M., Auerbach, A., and Breitman, M.: Dominant-negative and targeted null mutations in the endothelial receptor tyosine kinase, tek reveal a critical role in vasculogenesis of the embryo. Genes. Dev. 8:1897–1909, 1994.

Finn, M.C., Glowacki, J., and Mulliken, J.B.: Congenital vascular lesions: Clinical application of a new classification. J. Pediatr. Surg. 18:894, 1983.

Flye, M.W., Jordan, B.P., and Schwartz, M.Z.: Management of congenital arteriovenous malformations. Surgery 94:740, 1983.

Grabb, W.C., Dingman, R.O., O'Neal, R.M., and Dempsey, P.D.: Facial hamartomas in children: Neurofibroma, lymphangioma, and hemangioma. Plast. Reconstr. Surg. 66:509, 1980.

Gross, R.E.: Cystic hygroma. In *The Surgery of Infancy and Childhood.* Philadelphia: W.B. Saunders, 1953, pp. 960–970.

Haegerstrand, A., Dalsgaard, C.-J., Jonzon, B., Larsson, O., and Nilsson, J.:

Calcitonin gene-related peptide (CGRP) stimulates proliferation of human endothelial cells. Proc. Natl. Acad. Sci. U.S.A. 87:3299, 1990.

Jacobs, A.H., and Walton, R.G.: The incidence of birthmarks in the neonate. Pediatrics 58:218, 1976.

Kalischer, S.: Demonstration des Gehrins eines Kindes mit teleangiectasie der linksseitigen Gesichts-Kopfhaut und Hirnoberflache. Berl. Klin. Wochenschr. 34:1059, 1897.

Kennedy, T.L.: Cystic hygroma-lymphoma: A rare, still unclear entity. Laryngoscope 99 (49), 1–10, 1989.

Kirchoff, D., Eggert, H.R., and Agnoli, A.L.: Cavernous angiomas of the skull. Neurochirurgia 21:53, 1978.

Knowles, C.C.: Malocclusion associated with an extensive lymphangioma of the face. Trans. Br. Soc. Study Orthodont. 57:101, 1971.

Levine, C.: Primary disorders of the lymphatic vessels—A unified concept. J. Pediatr. Surg. 24:233, 1989.

Motly, R.J., Katugampola, G., and Lanigan S.W.: Videomicroscopy of vascular patterns in portwine stains predicts outcome. Laser Med. Surg., Abstr. 205(8), 37:1996.

Mulliken, J.B., and Glowacki, J.: Hemangiomas and vascular malformations in infants and children: a classification based on endothelial characteristics. Plast. Reconstr. Surg. 69:412, 1982.

Mulliken, J.B., and Young, A.E.: Vascular Birthmarks: Hemangiomas and Vascular Malformations. Philadelphia: W.B. Saunders, 1988.

Natali, J., Jue-Denis, P., Kieffer, E., et al.: Arteriovenous fistulae of the internal iliac vessels. J. Cardiovasc. Surg. 25:165, 1984.

Nilsson, J., Von Euler, A., and Dalsgaard, C.-J.: Stimulation of connective tissue cell growth by substance P and substance K. nature 315:61, 1985.

Ninh, T.N., and Ninh, T.X.: Cystic hygroma in children: Report of 126 cases. J. Pediatr. Surg. 9:191, 1974.

Oster, J., and Nielson, A.: Nucha naevi and interscapular telangiectasis. Acta Paediatr. Scand. 59:416, 1970.

Orten, S., Waner, M., Flock, S., Roberson, P., and Kincannon, J.: Port-wine stains: An assessment of 5 years of treatment. Arch. Otolaryngol. Head Neck Surg. 122:1174, 1996.

Phelps, C.D.: The pathogenesis of glaucoma in Sturge-Weber syndrome. Ophthalmology 85:276, 1978.

Pratt, A.G.: Birthmarks in infants. Arch. Dermatol. 67:302, 1967.

Rydy, M., Malm, M., Jernbeck, J., and Dalsgaard, C.: Ectatic blood vessels in port-wine stains lack innervation: Possible role in pathogenesis. Plast. Reconstr. Surg. 87:419, 1991.

Sabin, F.R.: The lymphatic system in human embryos, with a consideration of the morphology of the system as a whole. Am. J. Anat. 9:43, 1909.

Sako, Y., and Varco, R.: Arteriovenous fistula: Results of management of congenital and acquired forms, blood flow measurements, and observations on proximal arterial degeneration. Surgery 67:40, 1970.

Sato, T., Tozswa, Y., Deutson, U., Wolburg-Buchholz, K., Fujiware, Y., Gendron-Maguire, M., Godley, T., Wolburg, H., Risau, W., and Qin, Y.: Distinct roles of the receptor tyrosine kinases Tie-1 and Tie-2 in blood vessel formation. Nature 376:70–74, 1995.

Schmidt, G.H.: Hemangioma in the zygoma. Ann. Plast. Surg. 3:330, 1982.

Schnyder, U.W.: Zur Klinik and Histologie der Angiome. Arch. Dermatol. 200:483, 1955.

Smoller, B.R., and Rosen, S.: Port-wine stains: A disease of altered neural modulation of blood vessels. Arch. Dermatol. 122:177, 1986.

Stevenson, R.F., Thomson, H.G., and Morin, J.D.: Unrecognized ocular problems associated with "port wine" stain of the face in children. Can. Med. Assoc. J. 111:953, 1974.

Sturge, W.A.: A case of partial epilepsy, apparently due to a lesion of one of the vaso-motor centres of the brain. Trans. Clin. Soc. Lond. 12:162, 1879.

Vikkula, M., Boon, L., Carraway, K., Calvert, J., Diamount, A., Goumnerov, K., Pasyk, K., Marchuk, D., Warman, M., Cantley, L., Muliken, J., and Olsen, B.: Vascular dysmorphogenesis caused by an activating mutation in the receptor tyrosine kinase TIE2. Cell 87:1181, 1996.

Ward, G.E., Hendrick, J.W., and Chambers, R.G.: Cystic hygroma of the head and neck. Surg. Obstet. Gynecol. 58:41–47, 1950.

Waner M., and Suen JY.: Management of congenital vascular lesions of the head and neck. Oncology 9:989, 1995.

The Diagnosis of a Vascular Birthmark

MILTON WANER, M.D., F.C.S. (S.A.) AND
JAMES Y. SUEN, M.D., F.A.C.S.

Mulliken revolutionized the diagnosis of vascular lesions by providing us with a classification that is simple, easy to apply, and can be used to prognosticate with a high degree of accuracy. The most important distinction to be made is between a vascular malformation and a hemangioma. The reasons for this are twofold. From a prognostic standpoint, hemangiomas invariably involute, whereas vascular malformations hypertrophy. The former therefore may not need intervention, whereas the latter most certainly do. Furthermore, when considering the form of therapy, because hemangiomas involute, it may not be necessary to remove or destroy the entire lesion, but failure to erradicate the entire vascular malformation will invariably result in a recurrence. Once this distinction has been made, when dealing with a vascular malformation further refinement of the diagnosis will be needed. In

Hemangiomas and Vascular Malformations of the Head and Neck, Edited by Waner, M.D. and Suen, M.D.
ISBN 0471-17597-8 © 1999 Wiley-Liss, Inc.

the vast majority of cases, all that is needed to establish this diagnosis is a history and physical examination. Only rarely are special investigations necessary.

4.1. HISTORY

An accurate history will usually distinguish a hemangioma from a vascular malformation. To do this, three facts must be established:

4.1.1. Was the Lesion Present at Birth?

Although often helpful, the answer to this question may confuse the issue. Clearly, all vascular malformations are present at birth, but some, notably, arteriovenous malformations, may not be apparent until much later. Furthermore, venular malformations may not be noticed for several days and thus can be confused with an early hemangioma. On the other hand, while most hemangiomas are not present at birth, up to 30% may be. The answer to the second question is, therefore, of paramount importance.

4.1.2. Has it Grown?

An accurate account of the growth of the lesion (if any) is crucial. While both hemangiomas and vascular malformations "grow," only hemangiomas do so by hyperplasia (an increase in size due to cellular proliferation). Hyperplasia most often results in *rapid* growth, usually evident within the first few weeks of life. Although the rate and pattern of growth may vary between lesions, hyperplasia is only seen during the first year of life and ceases at some point between 8 and 14 months of age. Vascular malformations, on the other hand, expand by hypertrophy (an increase in size due to dilation of existing vascular channels). As one would expect, expansion by hypertrophy is a much *slower* process and takes place over many years. Furthermore, hypertrophy is relentless and will continue throughout the patient's life. The rate of expansion will be affected by several factors. High-flow lesions (arteriovenous malformations) will expand much more rapidly than low-flow lesions. Sepsis trauma and hormonal influences will also increase the rate of expansion.

4.1.3. Has it Shrunk?

This question is important when evaluating an older patient. Clearly, only hemangiomas involute, and do so at a variable rate, beginning after the first year of life. Vascular malformations never involute but do the opposite. They expand at a variable rate.

4.2. PHYSICAL EXAMINATION

By this stage of the encounter, a decision should have been made as to whether one is dealing with a vascular malformation or a hemangioma. Physical examination should not only confirm this, it should also take us one step further by providing important information as to the anatomical extent of the lesion. Where there is uncertainty, certain typical features may be helpful in arriving at a diagnosis.

4.2.1. Hemangiomas

The physical features of a hemangioma will be determined by two factors: its *depth* (with respect to skin) and its *stage* in its life cycle. With respect to depth, cutaneous and compound lesions will have an obvious, visible vascular component that will be bright red during proliferation and a dusky purple to greyish during involution. During the final phases of involution, evidence of epidermal atrophy, hypopigmentation, and telangiectasia are common. A subcutaneous lesion will merely present as a mass and may impart a bluish hue to the overlying skin. Once again, the deeper the lesion, the less obvious this discoloration.

During proliferation, the lesion will feel firm. An involuting lesion will feel softer.

4.2.2. Vascular Malformations

Several factors will determine the features of a vascular malformation, the most important of which are the vascular component and its depth with respect to skin. Because each type of malformation has its own unique features, it is convenient to describe them separately.

4.2.2.1. Venous Malformations

Venous malformations may be superficial or deep, localized or diffuse, unifocal or multifocal. Superficial lesions tend to expand toward a mucosal or cutaneous surface and eventually stretch the overlying membrane. This will impart a purple discoloration. With deeper lesions, the discoloration will be blue, but this will become less pronounced with increasing depth. Venous malformations are, as a rule, soft and compressible. They will, however, expand with raised venous pressure, and with deep palpation phleboliths may be felt. Unlike venular malformations (port-wine stains), venous malformations tend to be less uniform in distribution, but they do undergo nodule formation in their superficial areas with advancing age. The skin or mucosa overlying these is usually thin and atrophic, and bleeding is common with minor trauma.

Difficulty may be experienced with the diagnosis of a deep venous malformation. This will merely present as a mass with little or no discoloration of the overlying skin and may thus be confused with arteriovenous malformation or a subcutaneous hemangioma. An arteriovenous malformation will feel firmer and will not be compressible. A proliferating hemangioma will also feel firmer, but during involution a hemangioma may have a consistency similar to a venous malformation. A definite history of proliferation and/or involution will distinguish the two. Failure to do so will necessitate further investigation.

4.2.2.2. Venular Malformations

The diagnosis is usually unmistakable. Early pediatric port-wine stains are pink, macular, dermal vascular malformations involving one or more sensory dermatomes. These may on occasion be mistaken for early hemangiomas, but, once again, hemangiomas will invariably reveal their true nature by proliferating. With the passage of time, all port-wine stains will eventually thicken and darken in color, and most will form cobblestones. The mechanism for these changes is believed to be progressive ectasia of the venules, probably on the basis of a deficient or absent autonomic nerve supply. Mucosal involvement in continuity with the cutaneous component of a midfacial lesion is common. This will often result in hypertrophy of the involved structure. Thickening of the lip and/or increased protrusion of the maxilla are frequent examples.

An arteriovenous malformation or a venous malformation may involve the overlying skin, in which case the lesion may be mistaken for a venular malformation and an erroneous diagnosis may be made. More often than not, some degree of facial asymmetry will be evident and thus alert one as to the possibility of an underlying lesion. It is, therefore, of paramount importance to palpate all venular malformations. A firm mass or fullness is likely to be an arteriovenous malformation, whereas a soft compressible mass is probably an underlying venous malformation. Magnetic resonance imaging will, in these instances, offer a definitive diagnosis.

4.2.2.3. Midline Venular Malformations

Nevus flammeus neonatorum is a collective term used to describe true dermal venular malformations that are found on the nape of the neck (synonyms include salmon patch and stork bite) or in the midline of the face (also called angel's kiss). Other anatomical sites may also be involved, but these lesions are invariably midline. They are light pink macules that often fade within the first few years of life. Those that remain are unlikely to thicken or form cobblestones. Lesions involving the back of the neck are usually confluent, whereas those seen on the midline of the face are often nonconfluent or stippled.

4.2.2.4. Lymphatic Malformations

Lymphatic malformations may be localized or diffuse, superficial or deep, macrocystic or microcystic. A localized macrocystic lesion is most often found in the neck, whereas diffuse involvement is more apt to be microcystic and commonly involves the lower two-thirds of the face, mouth, and tongue. Diffuse involvement of the neck is also common along with facial involvement in the so-called cervicofacial variety. As the term implies, diffuse disease presents as widespread, brawny edema. Soft tissue, as well as bony hypertrophy, is common, and as a consequence facial features may be distorted and airway embarrassment is possible. Cutaneous or mucosal involvement manifests as multiple small vesicles filled with lymph. The depth of these vesicles can be quite deceptive, and they are usually connected with deeper cisterns. They may or may not co-exist with the presence of a deeper lesion.

4.2.2.5. Arteriovenous Malformations

Clinical features will vary according to the extent and the stage of the lesion as well as the age of the patient. Some confusion may arise as to whether this is indeed a congenital lesion because it may have only become apparent around the third, fourth, or even fifth decade of life. Closer questioning will often reveal that the lesion has, in fact, been present for some time prior. Arteriovenous malformations are congenital and therefore by definition are present at birth even though they may not be apparent until much later. The nidus of the arteriovenous malformation is usually palpable as a firm subcutaneous mass often warmer than the surrounding tissue. Contrary to widely held belief, it is usually not pulsatile. A thrill will be felt if the lesion is large enough, and a bruit may also be heard on auscultation. In the early stages, the hypertrophied feeding arteries and the dilated, tortuous veins draining the lesion may also not be apparent. However, with the passage of time, these will become evident. An overlying venular malformation may be present and thus mask the underlying arteriovenous malformation. All venular malformations should therefore be palpated. Rarely the phenomenon of stealing will result in flow reversal and necrosis of the overlying skin. Conversely, an increased blood supply may result in hypertrophy of the soft tissues and skeletal structures.

4.2.2.6. *Mixed Malformations*

A mixed malformation will assume the features of its subcomponents. A mixed venous–lymphatic malformation will consist of a diffuse subcutaneous mass of ectatic veins and lymph vessels and an overlying area of cutaneous or mucosal vesicles. The mass usually assumes a bluish hue, and many of the vesicles are filled with blood (instead of lymph). Mixed venous–venular malformations are also occasionally seen. Again, these will display features of both components. Hence, a venular malformation overlying, and often masking, a deeper venous malformation will be found. This underscores the importance of always examining venular malformations.

4.3. SPECIAL INVESTIGATIONS

A decision to proceed with special investigations should only be made if the diagnosis is in doubt or the outcome of the investigation will influence the decision to treat and/or the method of treatment. Because no single investigation is appropriate for all lesions, the benefits of a particular investigation should be weighed against its cost and morbidity.

4.3.1. Computerized Axial Tomography

Computerized tomographic (CT) scans have been replaced by magnetic resonance imaging (MRI) for the evaluation of soft tissue masses. Contrast-enhanced scans are still useful in evaluating intraosseus lesions as well as the bony margins of an extensive lesion under consideration for surgical resection. However, the need for intravenous contrast, the radiation exposure, and the occasional need for dynamic scanning are all disadvantages. This, coupled with the exquisite degree of sensitivity that MRI has to soft tissue details, has lessened the importance of a CT scan in the investigation of a vascular lesion. In spite of this, a CT scan will provide valuable diagnostic information. The extent of the lesion will be evident, and, with the use of intravenous contrast, one can distinguish between venous and lymphatic malformations. Dynamic scanning will even differentiate between high- and low-flow lesions. The risks associated with the higher dose of radiation used limit the use of this procedure.

4.3.2. Magnetic Resonance Imaging

An MRI is the most informative investigation in that it can determine the extent of the lesion with a high degree of accuracy as well as distinguish between the different types of malformation. More detail can be found in Chapter 6.

Hemangiomas are high-flow, solid tissue lesions of intermediate signal intensity on T1-weighted signal-enhanced presaturation images and high signal intensity on T2-weighted images. Hemangiomas enhance with gadopentate dimeglumine. Tubular flow voids are obvious.

Arteriovenous malformations are also high-flow lesions but lack a solid tissue structure and therefore display only intermediate signal intensity on T2-weighted signal-enhanced presaturation images. Flow voids are characteristically seen on T1- and T2-weighted signal-enhanced presaturation images and they lack significant enhancement on post-gadopentate T1-weighted images.

Venous malformations are low-flow lesions that exhibit low signal intensity on T1-weighted images, high signal intensity on T2-weighted images and exhibit variable enhancement with gadopentate dimeglumine.

Lymphatic malformations are also low-flow lesions that do not display high signal intensity on T1-weighted signal-enhanced presaturation images but do so on T2-weighted images. In other respects, these lesions display features not dissimilar to those seen with venous malformations. Macrocystic lesions will be obvious, and microcystic parenchymatous lesions will show a solid tissue mass of high signal intensity on T2-weighted presaturation images. Fluid–fluid levels are often seen especially in macrocystic lesions.

4.3.3. Angiography

The morbidity associated with angiography should preclude its use as a first-line investigation. When one considers the fact that an MRI can be used to diagnose the presence of an arteriovenous malformation, the diagnostic use of an angiogram should be restricted to mapping out the blood supply of the lesion. The only exception to this is the diagnostic dilemma associated with low grade arteriovenous malformations. Under these circumstances, the MRI features are not characteristic, and an angiogram is more likely to solve this problem.

An angiogram does, however, have considerable therapeutic value. Superselective embolization as a sole form of treatment or as pretreatment prior to surgical resection has become an integral part of the management of arteriovenous malformations.

4.3.4. Histology

Preoperative histological confirmation is seldom necessary, but the results of a permanent section are important in cases where a preoperative diagnosis was not made. Some confusion with regard to terminology is still common among pathologists, and the term *hemangioma* is still used generically to describe most vascular lesions. The detailed clinical pathological correlation in Chapter 5 will hopefully resolve this confusion.

4.3.4.1. Hemangiomas
The histological features of a hemangioma will depend on its stage of development. During the proliferative phase, the lesion is highly cellular and consists of tubules of plump, proliferating endothelial cells. Mitotic activity is high during active proliferation, and mast cells are

common. Vascular channels are not as obvious as one would expect, nor are red cells. As proliferation slows, so does the level of mitotic activity. It is important to realize that the progression from proliferation to involution is gradual. Pockets of proliferation may persist for some time, but eventually the scales will tip in favor of involution. As this happens, the pockets become less frequent. A diminution in the level of mitosis is accompanied by progressive flattening of the endothelial cells. Mast cells are now rare, and vascular channels filled with blood components predominate. The lesion thus assumes a more vascular appearance. The endothelial cell turnover rate eventually slows to normal, and, as this happens, the cells flatten and appear normal. The vessels eventually become replaced with fibro-fatty tissue.

4.3.4.2. Venous Malformations

These lesions are made up of a plexus of ectatic veins, varying in size from venules to large dilated cavernous veins. The vessel wall is made up of a single layer of flat endothelial cells, resting on a basement membrane and surrounded by a thin attenuated muscle layer.

4.3.4.3. Venular Malformations

Port-wine stains are characterized by ectatic postcapillary venules in the papillary and the mid-dermis. The degree of ectasia varies with the age of the patient and the maturity of the lesion. Although these two factors are related, the maturity and hence the degree of ectasia do not proceed at the same rate as the age of the patient. Recent evidence suggests that the progress of a venular malformation is related to the degree of absence of sympathetic venomotor tone (Smoller and Rosen, 1986; Rydh et al., 1990). Biopsies of infants reveal little, if any, abnormality apart from mild vascular ectasia (Barksy et al., 1980). As the lesion matures, the venules dilate and their walls become more attenuated.

4.3.4.4. Lymphatic Malformations

The histological features will depend on the clinical variant. The lesion is essentially made up of lymph-filled channels lined with a single layer of flat inactive endothelial cells on a basement membrane. The sizes of the cysts vary. In marcrocystic lesions large cysts dominate, whereas in microcystic lesions the opposite is true.

REFERENCES

Barsky, S.H., Rosen, S., Geer, D.E., and Noe, J.: The nature and evolution of port wine stains: A computer assisted study. J. Invest. Dermatol. 74:154, 1980.

Rydh, M., Malm, M., Jernbeck, J., and Dalsgaard, C.: Ectatic blood vessels in port-wine stains lack innervation: Possible role in pathogenesis. Plast. Reconstr. Surg. 87:419, 1991.

Smoller, B.R., and Rosen, S.: Port-wine stains: A disease of altered neural modulation of blood vessels: Arch. Dermat. 122:177, 1986.

5

The Surgical Pathology Approach to Pediatric Vascular Tumors and Anomalies

PAULA E. NORTH, M.D., PH.D. AND
MARTIN C. MIHM, JR., M.D.

Current clinical management of pediatric vascular lesions is usually based on clinical impressions, without the support of a standardized, reproducible histopathological diagnosis. This is due in large part to the fact that surgical pathologists, guided by the reference literature, have traditionally used the terms *hemangioma* and *angioma* indiscriminately to describe nonproliferative, developmental malformations as well as lesions that expand by true cellular hyperplasia. Commonly applied morphological modifiers such as "cavernous," "racemose," "simplex," and "capillary," many dating from the time of Virchow in the 1800s, often add more confusion than enlightenment. With this

Hemangiomas and Vascular Malformations of the Head and Neck, Edited by Waner, M.D. and Suen, M.D.
ISBN 0471-17597-8 © 1999 Wiley-Liss, Inc.

unfortunate usage of nomenclature, the pathologist often cannot reliably discriminate between lesions as clinically dissimilar as juvenile hemangiomas (which typically proliferate rapidly and then involute) and vascular malformations (true developmental disorders that do not proliferate or involute, but rather grow slowly with the child).

The ambiguities in classification of pediatric vascular lesions mirror our persistent incomprehension of the etiology of these lesions, which in turn is compounded by numerous published reports compromised by suspect histopathological classification. Despite these problems, which have been eloquently and thoroughly reviewed by Mulliken (1988), a growing number of immunohistochemical and ultrastructural studies of well-defined case compilations (Nichols et al., 1992; Mulliken and Glowacki, 1982; Kraling et al., 1996; Smoller and Rosen, 1986; Takahashi et al., 1994), complimented by in vitro and animal model studies of angiogenesis (Folkman, 1974; Folkman and Cotran, 1976; Risau et al., 1988; Zetter, 1988; Breier et al., 1992; Montesano et al., 1986; Mignatti et al., 1989), have suggested interesting possibilities. A standardized, reproducible histopathological classification scheme is clearly needed to support these investigative efforts, as has been discussed by many authors (Mulliken and Glowacki, 1982; Jackson et al., 1993; Coffin, 1997).

A useful plan for consistent histopathological classification of pediatric vascular lesions, congruent with clinical and radiological features, was presented in a study by Mulliken and Glowacki in 1982. According to this classification scheme, pediatric vascular lesions are first divided into two main categories: hemangiomas and vascular malformations. This division is based largely on the presence or absence of endothelial proliferation (visible mitotic figures or evidence of ^{3}H-thymidine incorporation). Hemangiomas, which show endothelial proliferation and a tendency to involute spontaneously, can then be subclassified according to the stage of their evolution (proliferative, involuting, involuted). Similarly, vascular malformations, which grow commensurately with the child, show little or no evidence of endothelial proliferation at any stage, and do not involute, can be subclassified as to the vessel type(s) represented in the lesion.

In this chapter, we adopt the straightforward classification of Mulliken and Glowacki (1982) and attempt to refine it for practicing surgical pathologists with provision of additional histological description that may prove useful in difficult cases. We also emphasize that in many cases consideration of the clinical history and time frame of the lesions (was it present at birth, did it proliferate, did it involute?) is essential. For example, an involuted hemangioma, with its ceased mitotic activity, flattened endothelium, vessel ectasia, and residual framework of feeding and draining vessels, may closely mimic a venous or arteriovenous malformation when considered out of context.

This chapter is limited by design to the pathology of lesions discussed in the clinical portions of this book: cutaneous and subcutaneous vascular lesions arising in the fetal and early postnatal periods, commonly referred to as *birthmarks*. Histologically similar lesions occur intracranially and in viscera and skeletal muscle (sometimes in combination with cutaneous lesions) and as part of a number of genetic syndromes. For these, the reader is referred to other published reviews (Burns et al., 1991; Spraker, 1986; Esterly, 1987; Coffin, 1997; Esterly, 1987).

5.1. NORMAL VASCULAR SYSTEM OF THE DERMIS AND SUBCUTIS

5.1.1. Structure

The vasculature of the dermis is composed of two major plexuses: a superficial one at the junction of the papillary and reticular dermis (Fig. 5.1A) and a deep one at the junction with the subcutaneous adipose tissue. These plexuses are fed by larger vessels that course through the subcutaneous and deep soft tissue planes. The superficial and deep plexuses each consist of extensively anastomosing networks of arterioles and venules and are joined by vertical channels and arcades of vessels traversing the reticular dermis. Capillaries loop upward from the superficial plexus into the papillae and carry blood from arteriole to venule, and vice versa (Mihm et al., 1976; Hood et al., 1993). The vessels of the deep plexus tend to be larger than the more superficial vessels and connect to the interlobular vasculature found within the fibrous septae of the underlying subcutaneous tissue. Lymphatics begin as collapsed, blind-ended channels in the superficial dermis and continue centrally into the subcutis. Cutaneous vessels are innervated by both sensory and sympathetic postganglionic nerve fibers (Bjorklund et al., 1986).

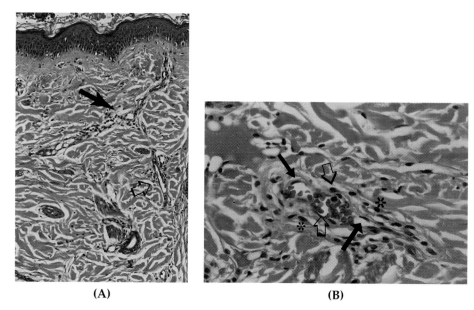

(A)   **(B)**

Figure **5.1**
Superficial dermal vasculature, normal skin. **(A)** Low-power view showing portions
of superficial plexus at junction of papillary and reticular dermis (solid arrow). A
small vertically oriented vessel (open arrow) joins the superficial plexus and courses
downward toward the deep plexus at the subcutaneous junction (out of view). **(B)**
High-power view of dermal capillary. Note elongated endothelial cells (solid arrows),
pericytes (open arrows), and perivascular mast cells (asterisks).

Capillary walls consist of a continuous single layer of endothelial
cells resting on a thin basement membrane that envelopes an incomplete layer of spindled pericytes (Fig. 5.1B). Small numbers of mast
cells and macrophages may be seen in the pericapillary connective tissue. The pericytes superficially resemble fibroblasts and provide the
only structural support for the capillary, as there is no investing fibrous
or smooth muscular tissue. Pericytes are closely related to vascular
smooth muscle cells by ultrastructure and composition of intermediate
filament proteins (Fujimoto and Singer, 1987). They immunoreact positively for smooth muscle cell actin–α and type IV collagen, but often
not for desmin. It is thought that pericytes, which may originate from
neural crest rather than mesoderm (Johnston, 1966; Nozue and Tsuzaki,
1974), can differentiate into smooth muscle cells of the media, and tran-
sitional cell types have been observed (Weber and Braun-Falco, 1973).
The basement membrane can be accentuated by periodic acid–Schiff
(PAS) staining and contains type IV collagen, laminin, and fibronectin.

Precapillary arterioles show rudiments of the three classic arterial layers: an intima consisting of endothelium anchored to a homogeneous basement membrane; a media consisting of a single, sometimes discontinuous layer of smooth muscle cells; and an outer, fibrous adventitia. An internal elastic membrane may or may not be visible in the smallest arterioles, although the thin media will show fine elastic fibers. Proceeding centrally, the arterial tree acquires progressively more smooth muscle, laced with fine elastic fibers, and the three coats become more distinct as the internal and external elastic membranes form.

Postcapillary venules resemble capillaries in that their walls consist solely of endothelial cells, basement membrane, and pericytes. The pericytes are more plentiful than those of capillaries, however, and the basement membrane may be multilayered. Unlike arterioles, venules do not have smooth muscle cells or elastic fibers. The transition to small veins is marked by the acquisition of a few smooth muscle cells, but no elastic tissue. The walls of larger veins do have elastic tissue and progressively more smooth muscle, but these supportive elements are fewer in number and more haphazardly arranged than those of arteries. Consequently, veins are often partially collapsed and irregular in shape in histological sections.

Small lymphatic vessels resemble capillaries and small venules at the light microscopic level, but by electron microscopy they can be seen to possess a scant, discontinuous basement membrane. There is some controversy as to whether the presence or absence of laminin immunoreactivity can be used to differentiate lymphatics from capillaries and small veins (Listrom and Fenoglio-Preiser, 1988; Hultberg and Svanholm, 1989). Larger lymphatics, like vascular channels, consistently immunoreact for laminin and may possess a scant smooth muscle coat.

5.1.2. Specialized Cellular Functions

Endothelial cells display an impressive assortment of synthetic, secretory, transport, and gatekeeping capabilities that are sensitive to the local cytokine microenvironment (for a review, see Cotran, 1989). Through the orchestrated synthesis and release of von Willebrand factor (vWf), factor V, plasminogen activator, and prostacyclin, and by the expression of thrombomodulin on its cell surface, the endothelium plays a central role in both coagulation and anticoagulation.

Endothelial cells can actively contract and can influence the growth of other cells in their vicinity through the secretion of growth factors such as basic fibroblast growth factor (bFGF) (Muthukrishnan et al., 1991). Interestingly, despite these varied, tightly controlled synthetic and secretory activities, most mature endothelial cells are thin and spindled shaped, with sparse free ribosomes and rough endoplasmic reticulum and an inconspicuous Golgi apparatus. As part of their gatekeeping function, they can be activated by the inflammatory mediators interleukin-1 (IL-1) and tumor necrosis factor-α (TNF-α) to express cell surface adhesive proteins such as endothelial leukocyte adhesion molecule 1 (ELAM-1) and E-selectin. These adhesive glycoproteins mediate adhesion of leukocytes to the endothelial surface in a multistep process, allowing migration of leukocytes into injured or antigen-stimulated tissue (Albelda et al., 1994). The endothelium of the postcapillary venule is the most active site for these adhesive interactions, and endothelial cells at this location are often more cuboidal and possess more structurally simple intercellular junctions than those elsewhere.

Pericytes, embedded within the basement membrane of capillaries and postcapillary venules, are active metabolic and regulatory elements in the vascular wall, in addition to providing a modicum of structural support. As a testament to their similarity to vascular smooth muscle cells, pericytes contain myosin and may be able to regulate blood flow by contraction (Joyce et al., 1985; Sims, 1986). They participate with the endothelium in synthesizing the basement membrane that envelopes them. They are capable of phagocytosis (Cancilla et al., 1972) and express both angiogenic mediators (vascular endothelial growth factor [VEGF] and urokinase) and inhibitors of angiogenesis (SMC-actin), tissue inhibitor of metalloproteinase (TIMP), and transforming growth factor-β (TGF-β) (Takahashi et al., 1994; Antonelli-Orlidge et al., 1989b).

Glomus bodies are specialized vascular structures found within the reticular dermis, especially in the skin of hands, feet, and ears, which form gated, direct shunts between arterioles and venules. Glomus bodies are heavily innervated by adrenergic nerve fibers, respond to local changes in temperature, and function in temperature regulation. Within each encapsulated glomus body, blood flow through one or more canals is controlled by encircling layers of epithelioid glomus cells. Glomus cells, like pericytes, contain contractile proteins and are closely related to smooth muscle cells (Venkatachalam and Greally, 1969; Miettinen et al., 1983).

5.1.3. Embryology

The vascular system begins to develop from embryonic mesoderm in the first weeks of gestational life, forming hematopoietic elements as well as extensively anastomosing endothelial tubes. Pericytes are recruited from the surrounding mesenchyme (or neural crest) to surround the tubes, and some of these eventually differentiate into the smooth muscle and fibroblastic elements of the media and adventitia. The plexiform networks of microvessels acquire arterial and venous stems at genetically determined locations and remodel to form the mature vascular system by selective growth and regression. Lymphatic channels develop as endothelial sacs that sprout from developing veins. The architectural maturation of the vascular system is normally complete by the end of the first trimester of fetal life. The specific molecular signaling mechanisms involved in this maturation process are not well understood. A number of endogenous agonists of angiogenesis, including VEGF, bFGF, platelet-derived growth factor (PDGF), TNF-α, transforming growth factor-α (TGF-α), and antagonists of angiogenesis, such as thrombospondin, interferons, and TGF-β, have been discovered in vitro and animal model systems. However, only two factors, acting through endothelial cell-specific receptor tyrosine kinases, have been shown to be critical in normal embryonic angiogenesis. First, VEGF, by binding to the Flt-1, Flt-4, and Flk-1/KDR family of receptor kinases, acts early in angiogenesis by influencing endothelial cell differentiation and proliferation (for a review, see Folkman and D'Amore, 1996). Second, a newly isolated factor, termed *angiopoietin-1*, by activating TIE1 and TIE2 receptors, appears to act later in angiogenesis by regulating the interaction of endothelial cells with surrounding mesenchymal components (Davis et al., 1996; Suri et al., 1996).

5.2. HEMANGIOMAS

These benign lesions are the most common tumors of infancy and characteristically evolve through three distinct stages: A rapidly proliferating stage (8–12 months) is followed by a prolonged involuting stage (1–12 years) and finally by a variably prominent end-stage fibro-fatty residuum that may or may not require surgical cosmetic correction. Female infants are three times more likely to develop hemangiomas

than males (Bowers et al., 1960). The suffix "-oma" in modern usage denotes a benign neoplasm and implies abnormal cellular hyperplasia.
Therefore, it is reasonable that the term *hemangioma* be applied only to
vascular lesions that proliferate and that show, at some stage in their evolution, histological evidence of endothelial mitotic activity. It is inappropriate and, in fact, misleading, to apply the term *hemangioma* to vascular
malformations because these are developmental anomalies that do not
proliferate, show little if any histological evidence of endothelial mitotic
activity, and do not involute. One of the most important roles of the
pathologist in this field is to distinguish between hemangiomas and vascular malformations, as the surgeon is occasionally mistaken in clinical
impression and may misjudge the likely biological behavior of the lesion.

The diagnosis *hemangioma* can be correctly applied, by the above ar-
guments, to only a handful of types of lesions. These lesions all display
benign endothelial and pericytic proliferation and include the classic
cutaneous or subcutaneous hemangiomas of infancy that involute (i.e.,
hemangiomas of the "usual type"), visceral hemangiomas comparable
in histology to cutaneous hemangiomas, perhaps a subcategory of the
"small vessel type" of skeletal muscle hemangiomas described by Allen
and Enzinger (1972), eruptive hemangiomas (pyogenic granulomas),
and tufted hemangiomas (tufted angiomas). The latter two entities
commonly proliferate in older children or adults and have sufficiently
distinctive clinical and pathological features to justify separate catego-
rization. Common adult-type hemangiomas ("cherry angiomas") have
been reported to demonstrate endothelial proliferation, although not
prominent, in very early lesions (Schnyder and Keller, 1954), although
the more mature lesions familiar to pathologists invariably show dilated capillary-sized vessels with a flattened, inactive-appearing endothelium.

Many pathologists prefer to label typical hemangiomas presenting
in infancy as "juvenile" or "infantile" hemangiomas. This practice, if
nothing else, serves to emphasize the clinical setting of these lesions
and does no harm. However, descriptive histological modifiers, such as
"capillary," "cellular," and especially "cavernous," should be avoided
because they are not only unhelpful, but have historically caused confusion with vascular malformations and with other, more aggressive
neoplasms that can also demonstrate these morphological features.
Juvenile hemangiomas are more meaningfully subclassified according

to stage (proliferating, involuting [with or without residual proliferative foci], and involuted or end stage) as described below. A microscopic description included in the surgical report can further clarify the pathological diagnosis to the clinician, who may have been confused by older terminology.

Multiple juvenile hemangiomas occur within two clinically distinct settings that vary more in distribution of lesions than histology: (1) *benign neonatal hemangiomatosis*, characterized by multiple cutaneous hemangiomas that involute (Held et al., 1990; Stern et al., 1981; Esterly et al., 1984); and (2) *diffuse* or *multiple neonatal hemangiomatosis*, characterized by multiple visceral and cutaneous hemangiomas, with high mortality (Holden and Alexander, 1970; Byard et al., 1991; Golitz et al., 1986; Gozal et al., 1990; Cooper and Bolande, 1965). Unfortunately, the general term *angiomatosis* has been used generically to encompass extensive vascular malformations as well as true hemangiomatoses and is stubbornly entrenched in the literature. Perhaps this, too, can change.

5.2.1. Histopathology

Hemangiomas in the early proliferative phase are characterized by well-defined, unencapsulated masses of plump, proliferating endothe-

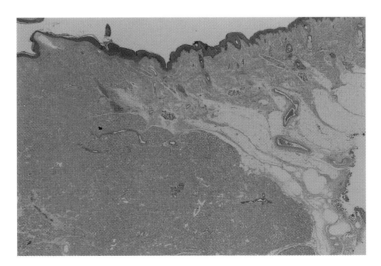

Figure **5.2**
Hemangioma, proliferative phase. This cellular lesion in an 11-month-old child has a well-defined border and is homogeneous throughout its depth.

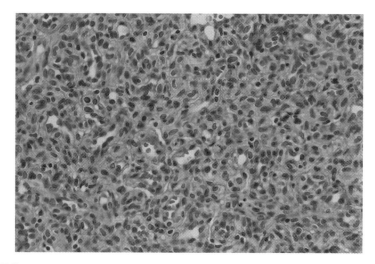

Figure 5.3
Hemangioma, proliferative phase. Small lumen formation is evident in this early
lesion from a 2-month-old infant.

lial cells and attendant pericytes that focally form small, rounded
lumina containing red cells (Figs. 5.2, 5.3). A reticulin stain will high-
light the rudimentary vessel formation by demonstrating thin reticulin
fibers encircling groups of endothelial cells (Fig. 5.4A). The organizing
endothelial tubes are invested by closely associated pericytes within a
PAS-positive basement membrane (Fig. 5.4B), without associated
smooth muscle cells. Even early hemangiomas show a tendency toward lobular arrangement of the proliferating cells and capillaries (Fig.
5.2, 5.5). Satellite nodules of proliferating capillaries may extend beyond the main tumor mass (Figs. 5.5B). Peripheral nerves intermingle
freely with the vascular elements, and granulated mast cells are numerous (Fig. 5.6). Endothelial and pericytic cells in proliferative phase hemangiomas show abundant, often clear cytoplasm and variably enlarged and hyperchromatic nuclei. Mitotic figures, normal in
configuration, are not hard to find and may be numerous, and immunoreactions for Ki-67 confirm that both pericytes and endothelial
cells are actively dividing (Fig. 5.7). Apoptotic bodies demonstrating
nuclear fragmentation are present in numbers roughly equivalent to
the number of mitotic figures (Fig. 5.8). These mitotic and apoptotic figures can be envisioned to represent opposing arms of a balance between growth and involution that shifts as the lesion matures.

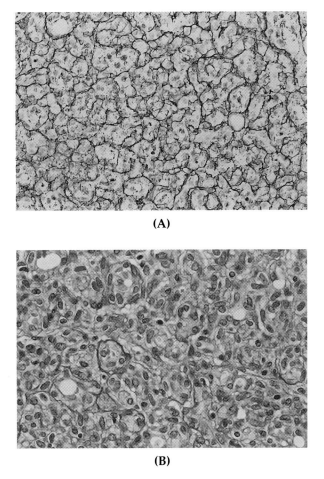

(A)

(B)

Figure 5.4
Hemangioma, proliferative phase. **(A)** Reticulin staining highlights individual vessels. **(B)** PAS staining accentuates basement membranes investing rudimentary vessels.

Anastomosing vascular channels, abnormal mitotic figures, and fascicles of spindle cells are not compatible with a diagnosis of hemangioma. Mitotic activity and mild nuclear pleomorphism, however, are integral features of the growing phase of these clinically benign vascular lesions and should not be cause for concern when present within the appropriate context of other histological features. These features, however, have prompted some authors to suggest the term *benign hemangioendothelioma* (Stout and Lattes, 1967; Anonymous, 1982) in an attempt to distinguish these benign lesions from the borderline

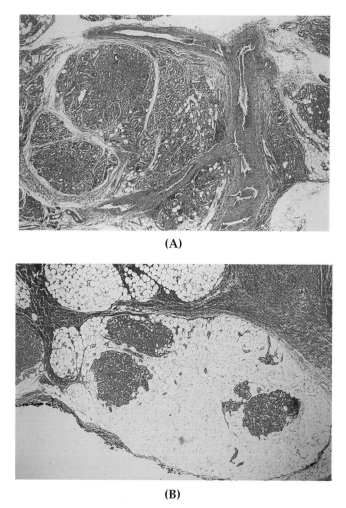

(A)

(B)

Figure 5.5
Hemangioma, proliferative phase, showing lobular architecture. **(A)** Lobules of proliferating capillaries, delineated by thin fibrous septi, surrounded by substantial arterial
feeder vessels. **(B)** "Satellite" lobules within adipose tissue adjacent to a persistently
proliferative lesion from a 31-month-old child.

malignant tumors known as *hemangioendotheliomas* and from angiosar-
comas (*malignant hemangioendotheliomas*). This nomenclature, although
well intended, is confusing and is not recommended.

Hemangiomas can be located superficially in the dermis, where
they impart a bright red color to the skin, or within the subcutaneous
tissue, where they appear blue or colorless, depending on depth. Many

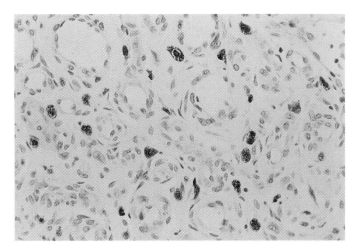

Figure 5.6
Hemangioma, early involuting phase, with persistent proliferative foci. Toluidine blue staining demonstrates increased density of mast cells (dark) surrounding lesional vessels.

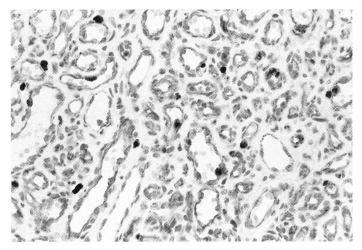

Figure 5.7
Hemangioma, proliferative phase. Immonoreaction for CD31 (light brown) is seen only within endothelial cells lining vascular lumina. Nuclear Ki-67 immunoreaction (dark brown) marks proliferative activity in both endothelial and perithelial cells.

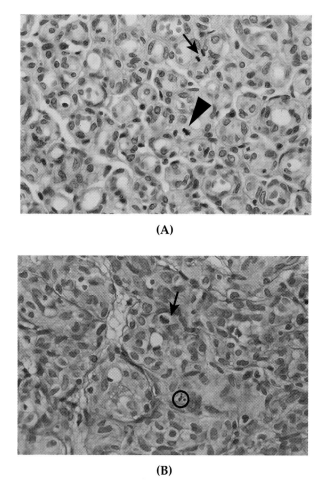

(A)

(B)

Figure 5.8
Hemangioma, proliferative phase. Note normally configured mitotic figures (arrows, **A, B**) and apoptotic figure demonstrating nuclear fragmentation (circle, **B**).

lesions extend from the superficial dermis into the underlying subcutaneous tissues. Clinicians have added their own confusion to that of the pathologists by referring to red superficial lesions as *capillary hemangiomas* and to the more blue subcutaneous lesions as *cavernous hemangiomas*. It is essential to recognize that *compound* hemangiomas involving both dermis and subcutaneous tissue are by microscopic examination largely homogeneous throughout their depth (Fig. 5.2). Thus, depth of involvement correlates with surface coloration, but not with size of the proliferating vessels.

The proliferating capillaries of superficial hemangiomas often involve contiguous skeletal muscle and salivary glandular tissue (Fig. 5.9). This is evident histologically not as an aggressive pushing border, but as insinuation of hemangioma elements between individual cells and glandular structures often without destruction of the overall architecture of these tissues. This gives a histological appearance of "having been there in the first place." Similarly, it is not uncommon to observe capillaries within the substance of nerves included in the lesion (Fig. 5.10) (Perrone, 1985; Calonje et al., 1995).

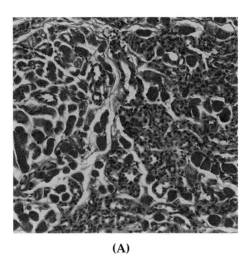

(A)

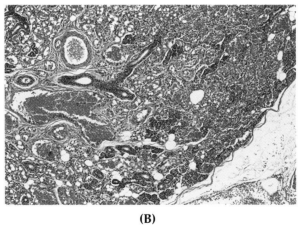

(B)

Figure **5.9**
Hemangioma, proliferative phase. These lesions may infiltrate **(A)** skeletal muscle or **(B)** salivary gland.

The boundaries of other structures surrounded by hemangiomatous elements may be surprisingly well preserved (Figs. 5.11–5.13).

Rapidly proliferating hemangiomas, like the histologically similar but grossly and clinically distinct pyogenic granuloma, may ulcerate, imposing secondary inflammatory changes on the histology of the lesion. Although this is a relatively common clinical event, it is not, in our experience, usually apparent in biopsied or resected specimens.

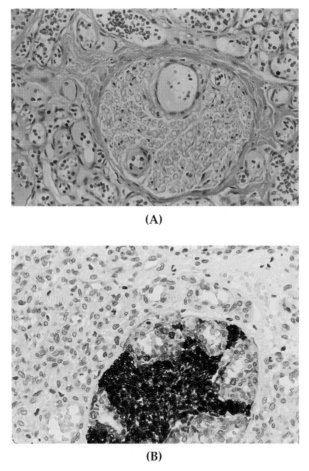

(A)

(B)

Figure **5.10**
Hemangioma, showing endoneural "invasion." Pseudosarcomatous endoneural invasion by hemangiomatous elements is evident in hematoxylin and eosin (H&E)-stained sections **(A)** and is highlighted by S-100 immunoreaction in which Schwann cells, but not blood vessels, are immunopositive **(B)**.

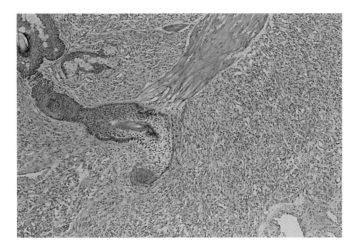

Figure 5.11
Hemangioma, proliferative phase. The proliferating capillaries closely abutt a dermal pilar unit, but do not infiltrate it.

Lesions demonstrating the localized, exophytic growth pattern of pyogenic granulomas (which most commonly occur in older children and adults on the face and oral mucocutaneous surfaces) can rarely present in infancy and have a pronounced tendency to become inflamed and bleed. Distinguishing histological features of pyogenic granulomas,

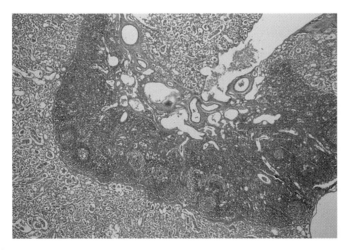

Figure 5.12
Hemangioma, early involuting phase. Hemangiomatous capillaries encircle, but do not infiltrate, a lymph node contained within a preauricular lesion. There is early maturation (ectasia) of component vessels.

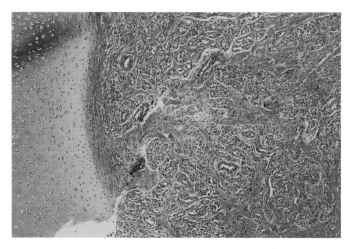

Figure 5.13
Hemangioma, proliferative phase. Lesion excised from the nasal tip of a 5-month-old infant to reduce the likelihood of deformity. The hemangiomatous vessels are excluded from the normally avascular nasal cartilage.

compared with juvenile hemangiomas, are an epidermal collarette (presumably resulting from rapid exophytic growth), reportedly normal mast cell density, secondary inflammatory changes (common), and, with maturation, thick fibrous septi (Fig. 5.14). It is unclear whether these exophytic lesions occurring in infants, perhaps best called *eruptive hemangiomas* (Marsch, 1981), are more closely related etiologically to pyogenic granulomas arising in patients or to the more classic *endophytic* juvenile hemangiomas. It is likely that all of these proliferative vascular lesions, which in their early stages share most histological and ultrastructural features (Marsch, 1981; Oehlschlaegel and Muller, 1964), are fundamentally related phenomena, modified by differing trigger mechanisms, local factors, or overall metabolic and hormonal environment.

Involution of hemangiomas becomes evident microscopically before the lesion begins to regress clinically. Apoptotic bodies and increased numbers of mast cells remain, while mitotic figures become less numerous. The endothelium begins to flatten, typically beginning at the periphery of the tumor, accompanied by lumenal enlargement (Fig. 5.15A). This maturation process begins at different times in differ-

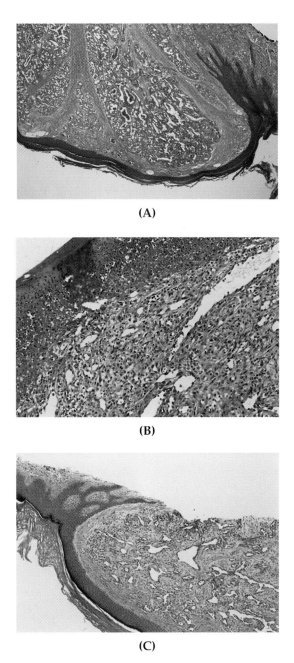

(A)

(B)

(C)

Figure **5.14**

Pyogenic granulomas in infants. **(A)** Exophytic, localized proliferation of vascular elements, similar in histology to juvenile hemangioma. Note epidermal collarette and thick fibrous septi, which are characteristic of this clinically distinct variant of hemangioma that is most common in older children and adults. **(B)** Ulceration is common in pyogenic granulomas due to their rapid, exophytic growth pattern and susceptibility to trauma. Note histological similarity of the proliferating vascular elements to juvenile hemangioma. **(C)** This more mature lesion shows decreased cellularity and a more fibrotic stroma. Note epidermal collarette and hyperkeratosis.

ent portions of the tumor (Fig. 5.15B) and lends a "mixed capillary–cavernous" look to the tumors (Fig. 5.16). Such morphological qualifiers, however, are confusing and, as discussed previously, should not be used. During early involution, component vessels remain markedly increased in number, and the dilating capillaries are closely packed, with little intervening connective tissue stroma (Fig. 5.17). Mitotic figures may still be seen despite flattening of the endothelium (Fig. 5.17B). As involution proceeds, the once-proliferating vessels decrease in number, and loose fibrous or fibroadipose tissue begins to separate vessels both within and between lobules (Figs. 5.18, 5.19). Lobules with persis-

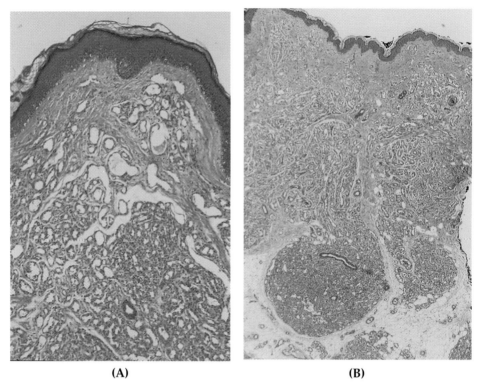

(A)   (B)

Figure 5.15
Hemangioma, early involuting phase. **(A)** Early ectasia of lesional vessels and flattening of endothelium, most prominent in the superficial portions of the lesion. Vessel drop-out is minimal. **(B)** In this lesion, involution is more advanced in the dermal lobules than in those in the underlying adipose tissue.

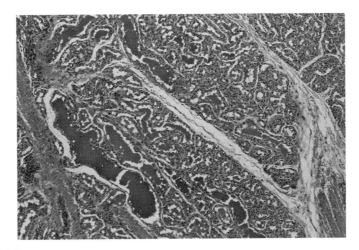

Figure 5.16
Hemangioma, early involuting phase. Uneven dilation of vessels during early involution may impart a mixed "capillary–cavernous" appearance.

tently proliferating endothelial cells and small lumina become less numerous as the connective tissue replacement progresses. Hemangiomas
throughout the involuting phase continue to show an abnormally high,
albeit decreasing, vessel density and may be subclassified usefully by
the pathologist as to whether residual proliferative (mitotically active)
foci remain. The lesion may resemble a venous or arteriovenous malformation when mitotic activity becomes inapparent (Fig. 5.20). In end-
stage lesions (Figs. 5.21, 5.22), all that remains is a fibro-fatty background with a mast cell count comparable to that of normal skin,
studded by a few residual vessels similar to normal capillaries or
venules and scattered larger vessels with fibrotic walls that presumably
serviced the original lesion (see discussion below). No endothelial or
pericytic mitotic activity remains. Lesions complicated by repeated ulceration that focally have destroyed the papillary dermis will show epidermal atrophy and underlying fibrous scar tissue, with loss of dermal
appendages (Fig. 5.23). Foci of dystrophic calcification will rarely be
seen. Lesions that fail to involute completely will continue to show islands of proliferating capillaries, with other areas showing vessel drop-
out and fibro-fatty infiltration (Fig. 5.24).

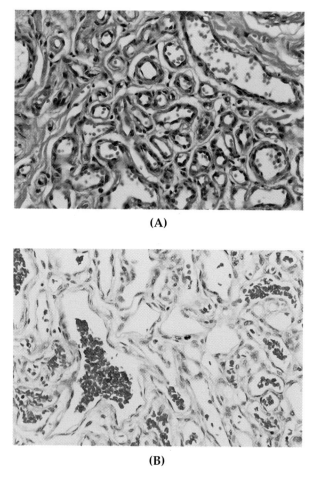

(A)

(B)

Figure 5.17

Hemangioma, involuting phase. **(A)** Note flattening of endothelium and tight alignment of pericytes around vessels. Poorly organized masses of plump, immature pericytes and endothelial cells are not seen, but vessel density remains high. Mitotic activity is not evident in this example. **(B)** This lesion from a 2-year-old child shows persistent mitotic activity despite considerable maturation of vessels.

The mass of feeding and draining vessels seen within hemangiomas is sometimes alarming, demonstrating thickened, often asymmetrical walls reminiscent of a vascular malformation (Figs. 5.25–5.27). These vessels are present even in very young lesions, but become more obvious during involution and many persist in end-stage lesions. Their apparently increasing prominence during late involution is probably, at least in part, an illusion, caused by disappearance of the intervening

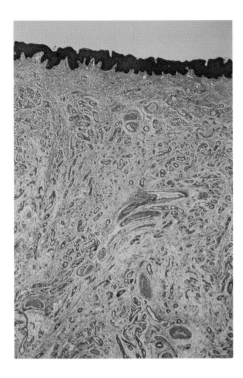

Figure 5.18
Hemangioma, late involuting phase. Fibro-fatty tissue has replaced many lesional vessels, although the vessel density remains abnormally high. Feeder vessels with thicker walls are also present. These presumably serviced the earlier, larger lesion.

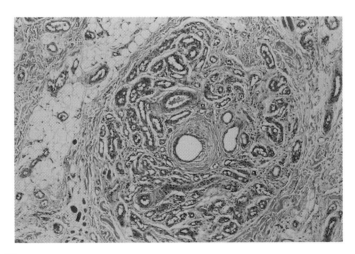

Figure 5.19
Hemangioma, late involuting phase. A preserved lobule of maturing hemangiomatous vessels surrounds larger feeding and draining vessels in this late involuting lesion from a 2-year-old child.

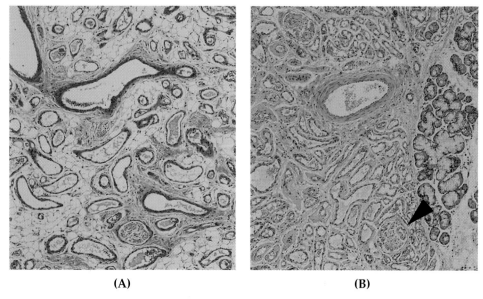

(A)   (B)

Figure 5.20
Hemangiomas, late involuting phase. High-power views of late involuting hemangiomas from subcutaneous tissue **(A)** and salivary gland **(B)**. Vessels remain high in density, but are separated by loose connective tissue. Endothelia are flat, and no mitotic activity is evident. This stage of hemangioma may be mistaken for venous malformation without careful histological examination and clinical correlation. Note vascularized nerve in B (arrow), a feature common to hemangiomas.

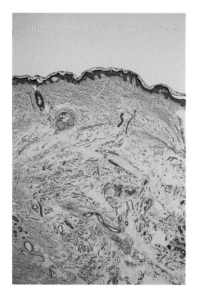

Figure 5.21
Hemangioma, involuted (end stage). Clinically stable lesion from a 6-year-old child demonstrating a fibro-fatty residuum studded by a few persistent component and supportive vessels without proliferative activity. Although a small degree of further vessel drop-out might have occurred in this lesion over the next several years, the lesion is essentially end stage histologically and was excised for cosmetic purposes.

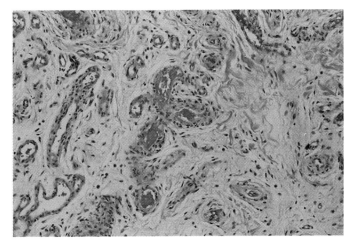

Figure 5.22
Hemangioma, involuted (end stage). High-power view of an involuted hemangioma showing sparse residual vessels, similar to normal capillaries and venules,
within a loose fibrous stroma. Electron microscopy would reveal a multilaminated
basement membrane surrounding the persistent vessels, reflecting past cycles of
endothelial and pericytic cell turnover (see text).

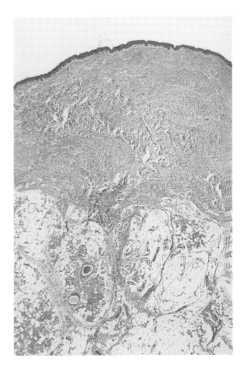

Figure 5.23
Hemangioma, late involuting, scar tissue. Note epidermal atrophy and dermal scar tissue (with loss of appendages) overlying this involuting
hemangioma complicated by repeated
ulceration.

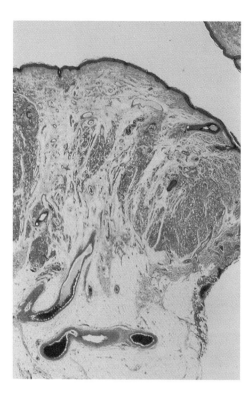

Figure 5.24
Hemangioma, focal involution with persistent proliferative lobules. Vessel drop-out and fibro-fatty replacement are advanced in the central portion of this lesion, while cellular proliferative lobules of the lesion persist. Note large feeder vessels in the subcutaneous tissue underlying the lesion.

masses of capillaries. However, to our knowledge no quantitative studies of this phenomenon are available. These larger vessels within hemangiomas presumably hypertrophy in response to shunting of blood flow through the proliferating capillaries as they form lumina and dilate. Some may originate from the proliferating vessels themselves, thickened by perivascular fibrosis. These "secondary" vascular malformations, if you will, are clinically silent after involution of the capillary bed, and, unlike true "primary" vascular malformations, they do not continue to grow with the child. This suggests that they arise through an intrinsically different pathogenetic mechanism. For the pathologist examining a specimen, the problem is simply one of recognizing the difference between the residual large-vessel framework of an involuted hemangioma and a true vascular malformation. This can usually be accomplished by first considering the clinical history and overall histological appearance of the lesion and by judging the degree of mural atypia of the component vessel walls.

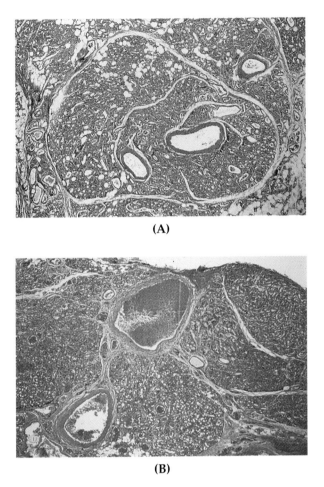

(A)

(B)

Figure **5.25**
Proliferative hemangiomas showing supportive vessels. **(A)** Large caliber, but anatomically normal feeding and draining vessels servicing lobules of a persistently proliferative hemangioma in a 2-year-old child. **(B)** This hemangioma from a 3-year-old patient exhibited high flow and was clinically mistaken for an arteriovenous malformation. Enlarged, thick-walled veins suggest significant secondary arteriovenous shunting through the hemangiomatous capillary bed. Mitotic foci were still present (not shown).

Immunohistochemical studies of hemangiomas reveal a mixture of endothelial cells and pericytes, with lesser numbers of mast cells, fibroblasts, and interstitital cells that express factor XIIIa (Gonzalez-Crussi and Reyes-Mugica, 1991; Smoller and Apfelberg, 1993; Pasyk et al., 1982). In general, the endothelial cells demonstrate the immunophenotype of normal, mature endothelium, immunoreacting positively for CD31,

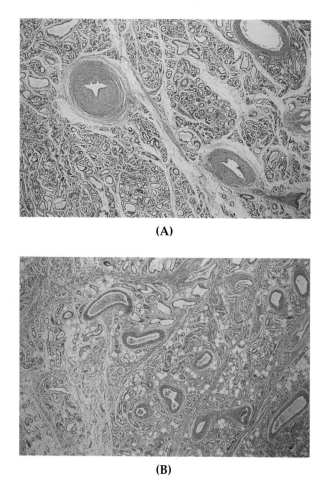

(A)

(B)

Figure 5.26
Involuting hemangiomas showing supportive vessels. Note the prominent residual framework of thick-walled veins and arteries in these two examples of involuting hemangioma (A, B). Hemangiomas at this stage may be misinterpreted as arteriovenous malformations.

CD34, factor VIII–related antigen (vWf), *Ulex europaeus* lectin I, VE-cadherin, HLA-DR, and vimentin (Figs. 5.28–5.30) (Martin-Padura et al., 1995; Gonzalez-Crussi and Reyes-Mugica, 1991; Smoller and Apfelberg, 1993; Takahashi et al., 1994; Suzuki et al., 1986). However, cellular, solid-appearing areas of hemangiomas, where mitotic activity is the highest, stain more weakly for vWf than do maturing areas with distinct vessel formation. This is probably due to dilution of stored vWf in endothelial

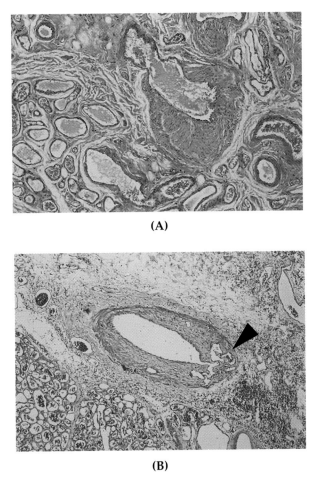

(A)

(B)

Figure 5.27

Atypical veins in involuting hemangiomas. **(A)** A vein with an irregular smooth muscle coat in an involuting hemangioma. **(B)** A vein showing thin, fragmented media (arrow) and thickened intima with focal recanalization. Adjacent hemangiomatous capillaries show ectasia and endothelial maturation.

cells by repeated cycles of mitosis and is accompanied by a decrease in number of vWf-containing Weibel-Palade bodies per cell (Alles, 1987). Thus, immunohistochemical studies of very immature hemangiomas may be limited by this immunohistochemical insensitivity and heterogeneity (Yasunaga et al., 1989). Recently, Kraling and coworkers (1996) reported increased expression of E-selectin, an endothelial cell-specific

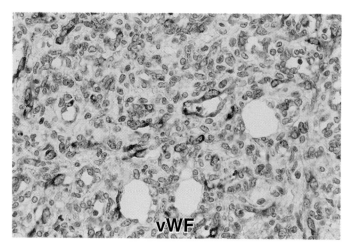

Figure 5.28
Hemangioma, proliferative phase. Immunoreaction for vWf is positive in endothelial cells of better-differentiated vessels and in scattered cells within more solid areas. The lesion is from a 3-month-old infant.

leukocyte adhesion molecule, in proliferative phase hemangiomas compared with involutive phase hemangiomas and quiescent endothelium, suggesting that E-selectin is a marker for proliferative endothelium and supporting assertions that it functions in angiogenesis (Nguyen et al., 1993; Koch et al., 1995).

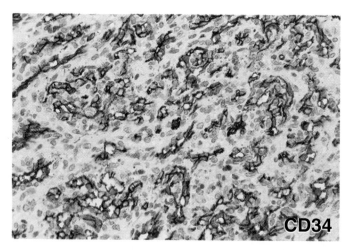

Figure 5.29
Hemangioma, proliferative phase. Immunoreaction for CD34 is positive in endothelial cells lining small lumina, while intervening cells, primarily pericytes, are not reactive. The lesion is from a 1-month-old infant.

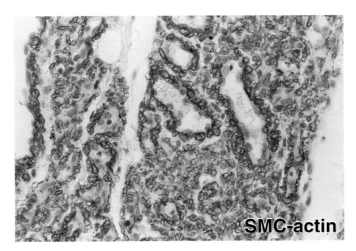

Figure 5.30
Hemangioma, proliferative phase. Immunoreaction for smooth muscle cell actin-α (SMC-actin) is seen in perithelial cells encircling vessels and in many cells within intervening solid areas. Endothelial cells lining vessel lumina are not reactive. The lesion is same as that shown in Figure 5.28.

Electron microscopic examinations of proliferative phase hemangiomas show small lumina formed by plump endothelial cells resting on a multilaminated basement membrane enveloping a cuff of pericytes (Mulliken and Glowacki, 1982; Waldo et al., 1977; Gonzalez-Crussi and Reyes-Mugica, 1991; Pasyk et al., 1982). Mast cells align their long microvillous projections parallel to the laminations of the basement membrane and form close contacts with fibroblasts and macrophages (Dethlefsen et al., 1986). The endothelial cells contain active-appearing rough endoplasmic reticulum and occasional Weibel-Palade bodies, which may be difficult to find. Lamellar crystalline inclusions, similar to those seen in fetal endothelium, have been described in the endothelial cytoplasm of hemangiomas and are presumably an indication of immaturity (Kumakiri et al., 1983; Pasyk et al., 1983). Multiple, redundant layers of basement membrane are seen in immature hemangiomas, persist in a thinned, fragmented form even in end-stage lesions, and may result from repeated cycles of cell division and cell death (Mulliken and Glowacki, 1982). The vessels comprising vascular malformations, which do not manifest such endothelial proliferation, possess only single-layered basement membranes (Mulliken and Glowacki, 1982).

5.2.2. Pathogenesis

Any discussion of the pathogenesis of hemangiomas must be prefaced by the statement that we still understand very little about the etiology of these lesions. However, the potential usefulness of hemangiomas as models of angiogenesis has spawned considerable recent interest in hemangiomas, and useful findings are beginning to emerge. We present here a discussion of current theories and the available evidence, or in some cases lack of evidence, for these theories.

The conceptualization of hemangiomas as benign, clonal neoplastic proliferations is simple and appealing. At present, however, conclusive genetic evidence for clonal expansion of endothelial or perithelial cells in hemangiomas is lacking. Clonality analyses based on differential methylation of active versus inactive X-linked alleles, such as those for the human androgen receptor or glucose-6-phosphate dehydrogenase genes in female patients heterozygous at these sites, has been successfully applied to many types of tumors. Unfortunately, the intrinsic cellular heterogeneity of hemangiomas has limited the usefulness of this type of analysis for these lesions. Cell-selective microdissection techniques may overcome this problem in the future.

Alternative etiological theories, based not on neoplasia but on focal persistence of fetal angioblastic tissue or local arrest in a primitive stage of vascular morphogenesis, have been proposed for many years (de Takats, 1932; Watson and McCarthy, 1940; Pack and Miller, 1950; Malan, 1974; Kaplan, 1983; Stal et al., 1986). While these developmental theories have some intuitive appeal, they have not provided insight into specific mechanisms of pathogenesis. Some investigators have conceptualized the capillary proliferation in hemangiomas as a reaction to aberrant local release of angiogenic factors, analogous to the abnormal capillary proliferations characterizing retrolental fibroplasia, seen in premature infants after receiving oxygen therapy, and diabetic retinopathy (Reese and Blodi, 1951; Andrews and Domonkos, 1953; Folkman and Klagsburn, 1987). Such theories might explain the reported observation that hemangiomas occur up to twice as often in preterm infants as in full-term babies (Amir et al., 1986; Powell et al., 1987; Mulliken, 1988).

Our understanding of the local cytokine microenvironment in which hemangiomas evolve has been significantly advanced by recent immunohistochemical studies. Takahashi et al. (1994) compared the density of vessels immunoreacting for a gauntlet of markers in proliferating, involuting, and involuted hemangiomas. They found high levels of expression for proliferating cell nuclear antigen (PCNA), VEGF, and type IV collagenase in the proliferating phase and high levels of expression within pericytes of TIMP and SMC-actin, as well as maximal mast cell counts, in the actively involuting phase. Expression of bFGF, one of the most potent angiogenic factors known, was significantly higher in the proliferating phase than in the involuted phase. In contrast, and not surprisingly, vascular malformations manifested low or absent expression of all of these markers, as did hemangiomas in the fully involuted stage. Based on this evidence, Takahashi and coworkers (1994) proposed that endothelial and perithelial mitotic activity in hemangiomas is mediated in part by the angiogenic peptides VEGF and bFGF, with type IV collagenase and urokinase clearing space for expansion. Differentiating perithelial cells might then express TIMP, which might, in turn, promote involution by inhibiting the angiogenic proteases. Mast cells might contribute to the involution process by secreting inhibitors of angiogenesis such as interferon-γ and TGF-β (Gordon et al., 1990; Freisel et al., 1987) and by stimulating fibrosis. A clear understanding of the complex interplay of molecular and cellular factors underlying hemangioma growth and involution, provided by studies such as these, should enable us to better evaluate mechanisms of action and to better predict efficacies of newly proposed angiostatic drug therapies.

The potential role of pericytes in the evolution of hemangiomas deserves close attention because these cells, as well as endothelial cells, contribute significantly to the cellular proliferation seen in growing hemangiomas. In a study of co-cultures of endothelial cells and pericytes, Antonelli-Orlidge et al. (1989a) found that pericytes secrete TGF-β, a known inhibitor of endothelial cell proliferation, but that the secreted TGF-β is activated from its latent form only when endothelial cells and pericytes are in close contact with one another. The relevance of these findings in culture to hemangiomas in vivo is, of course, not clear, but they raise the intriguing possibility that disruption in normal pericyte–endothelial cell interactions, perhaps by simple overgrowth, might "take the brakes off" endothelial cell proliferation. This is worthy of continued study. Recently, Ito and colleagues (1997) reported an increased circulating level of TGF-β1 in a patient with a large hepatic hemangioma and demonstrated overexpression of the TGF-β1 gene in the hemangioma tissue. The high circulating level of TGF-β1 was associated with transient immunosuppression in this patient and resolved after surgical removal of the tumor. Further evidence for the importance of stromal cells in the pathogenesis of hemangiomas has recently been presented by Berard et al. (1997). They found that stromal cells cultured from juvenile hemangiomas release VEGF and elicit an angiogenic response blocked by neutralizing anti-VEGF IgG when grafted into nude mice.

The possible role of mast cells in the evolution of hemangiomas also has received much attention because of the notable abundance of mast cells in growing and involuting hemangiomas as well as in a number of other highly vascularized tumors (for review, see Meininger and Zetter, 1992). Even in normal tissues mast cells accumulate near vascular structures. Glowacki and Mulliken (1982) reported 30–40-fold more mast cells within proliferating hemangiomas than in normal-aged and site-matched tissue, whereas vascular malformations and involuted hemangiomas had normal mast cell counts. Other quantitative studies have shown more modest, but still significant increases (Pasyk et al., 1984; Takahashi et al., 1994). Meninger et al. (1995) have observed increased release of stem cell factor from hemangioma-derived endothelial cells in tissue culture compared with release from cells derived from normal endothelium. As stem cell factor is a principle cytokine regulating the growth and differentiation of mast cells (Tsai et al., 1991), acting in part through suppression of mast cell apoptosis (Iemura et al., 1994; Mekori et al., 1993), this may explain the mast cell accumulation seen in hemangiomas.

The role of mast cells in hemangiomas is unlikely to be simple because these complex and heterogeneous cells are capable of releasing a myriad of cytokines that include both agonists and antagonists of angiogensis and mediators of fibrosis (for review, see Gordon et al., 1990). One of these cytokines, bFGF, is a potent stimulator of both endothelial and fibroblastic proliferation and has been localized to mast cells residing within hemangiomas (Qu et al., 1995). In addition to the association between mast cells and proliferating vascular elements in hemangiomas, mast cell accumulations are associated with the neovascularization of wound repair and diabetes mellitus, of chronic inflammatory processes such as rheumatoid arthritis and Crohn's disease, and of animal models of angiogenesis (Smith and Basu, 1970; Kessler et al., 1976; Norrby et al., 1989). Increased mast cells have also been observed in a number of chronic fibrotic states, including scleroderma and idiopathic pulmonary fibrosis (Hunt et al., 1992; Hawkins et al., 1985; Claman, 1989). Pasyk and colleagues (1984) have noted the coincident appearance of increased mast cell counts and fibrous tissue in growing hemangiomas and have suggested that mast cells contribute to fibrous involution rather than to endothelial proliferation in these lesions. The previously discussed finding by Takahashi's group that mast cell numbers are highest in the early to middle involuting phase of hemangiomas is consistent with this idea. Clearly we are just beginning to understand the role of mast cells in hemangioma growth and involution. It seems likely that this role is facilitative rather than directive.

Rarely, lobules of proliferating, mitotically active capillaries suggestive of hemangioma are found within an otherwise typical vascular malformation. We have observed occasional cases such as these in our own files, and there are a few reports in the literature of lobular hemangiomas clinically thought to be pyogenic granulomas, as well as one reported case of a tufted angioma, arising within pre-existing port-wine stains (Swerlick and Cooper, 1983; Barter et al., 1963; Warner and Wilson-Jones, 1968; Alessi et al., 1986). In addition, hemangiomas with persistent high blood flow have been described in which a labyrinth of proliferating hemangiomatous capillaries seem to have "tapped into" a large caliber arterial supply, forming a significant physiological shunt (Martinez-Perez et al., 1995). These observed associations have supported speculations that altered blood low or local cytokine microenvironment within congenital vascular malformations might stimulate proliferation of endothelial cells derived from the abnormal vessels. This idea remains speculative, however, and awaits experimental support.

5.3. VASCULAR MALFORMATIONS

Vascular malformations are true anomalies of embryonic vascular morphogenesis and thus by definition are always present at birth. The great majority of those involving skin and subcutis are not only present but clinically obvious at birth. A small number, however, like their more deeply seated visceral, intramuscular, and intracranial counterparts, may not become evident until adolescence or even adulthood. Because vascular malformations result from developmental error rather than abnormal cellular proliferation, their clinical behavior is quite distinct from that of hemangiomas. Malformations generally demonstrate slow growth, proportional to that of the child, and do not involute. Occasional periods of rapid enlargement, most commonly seen in puberty or pregnancy, or as the result of trauma, are attributable to hemodynamic factors such as thrombosis, progressive ectasia, or formation of new arteriovenous communications, and not cellular hyperplasia. The histological correlate of this behavior is a flat, inactive-appearing endothelium in lesions of all ages.

Vascular malformations may contain any combination of lymphatic, capillary, and venous components, reflecting the integrated development of the various parts of the vascular system during embryonic development. Every possible combination has been reported, and complex mixed malformations have been associated with a number of dysmorphic syndromes (for review, see Burns et al., 1991). In addition, many large vessel malformations are associated with an overlying cutaneous venulocapillary malformation (similar to a port-wine stain). In the later instance, the clinical behavior of the lesion is determined by the underlying large vessel component. It is useful to divide mixed malformations into high-flow and low-flow lesions, a distinction made in correlation with clinical and radiological findings. Arteriovenous shunting through a malformation produces a high-flow lesion that may cause clinical complications such as congestive heart failure, ischemic necrosis of adjacent skin due to arterial "steal," destruction of adjacent bone, and life-threatening hemorrhage. The natural history and therapy of the various clinically recognized subtypes of vascular malformations are presented in Chapters 3 and 10. Radiological assessment of vascular malformations, which becomes particularly important in evaluating arteriovenous malformations, is presented in Chapter 6. We describe here the histological features of the clinically recognized subtypes of these lesions and discuss possible pathogenic mechanisms.

5.3.1. Lymphatic Malformations

Lymphatic malformations have traditionally been referred to as *lymphangiomas* despite an absence of demonstrable endothelial mitotic activity and represent the most common type of vascular malformation presenting in children and adolescents (Coffin and Dehner, 1993). Most, if not all, "lymphangiomas" are true malformations, although some controversy persists as to whether rare lesions may have proliferative potential (Bowman et al., 1984). Lymphatic malformations can be limited to the dermis and epidermis or may extend into deeper soft tissues, viscera, or bone. Acquired lesions of similar histological appearance sometimes occur in adults and are referred to as *lymphangiectasias* (Prioleau and Santa Cruz, 1978).

Lymphatic malformations located in the dermis or submucosa consist of vessels filled with clear fluid, and sometimes a few lymphocytes or even erythrocytes, lined only by a single layer of flattened endothelial cells. These cells immunoreact most reliably for vWF and CD31 (Ramani and Shah, 1993). Strong immunoreaction for laminin has also been reported (Autio-Harmainen et al., 1988). Ectatic lymphatic vessels located in the papillary dermis or superficial submucosa appear grossly as multiple cutaneous or mucosal blebs. Lesions of this type have been referred to as *lymphangioma simplex* (Wegner, 1877) or *lymphangioma circumscriptum* (Morris, 1889) for over a century. The overlying epidermis of such lesions shows variable acanthosis and hyperkeratosis, and the surrounding dermis may show a sparse lymphocytic infiltrate (Figs. 5.31, 5.32).

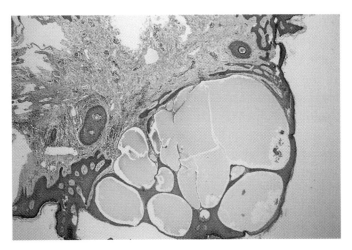

Figure **5.31**
Cutaneous lymphatic malformation ("lymphangioma circumscriptum"). Dilated lymphatic channels in papillary dermis form a large subepidermal bleb. Note focal dermal chronic inflammation.

Figure 5.32
Cutaneous lymphatic malformation ("lymphangioma circumscriptum"). Dilated lymphatic spaces expand a dermal papilla. Dilated lymphatic channels are also evident in the adjoining reticular dermis and are associated with a large vessel lymphatic malformation in the underlying soft tissue (not shown). A similar type of cutaneous lesion, characterized by hyperkeratosis and markedly dilated, blood-filled spaces in the papillary dermis ("angiokeratoma circumscriptum"), may overlie a deep venous malformation.

Lymphatic malformations located more deeply within subcutaneous tissues, traditionally referred to as *cavernous lymphangiomas*, consist of larger, more irregular lymphatic vessels, including some with very thin, incomplete smooth muscular coats (Fig. 5.33). These vessels, like those in the dermal lesions, are filled with proteinaceous fluid and lymphocytes, and sometimes erythrocytes, and may have valves. The surrounding loose connective tissue stroma contains a variable lymphocytic infiltrate, often with prominent lymphoid follicle formation. The latter feature is helpful in distinguishing lymphatic malformations from venous malformations, which do not share this marked lymphopoietic component. Dilated lymphatics may extend between bundles of underlying skeletal muscle and produce a multiloculated cyst that often appears to be well circumscribed by a thin fibrous "capsule." Infection and inflammation are common in these lesions, presumably due to ease of entry provided to bacteria by thin, traumatized lymphatic vessels, and these secondary phenomena are responsible for the surrounding rim of fibrosis. This apparent encapsulation may be deceptive in defining the outer extent of the malformation. Surgical mar-

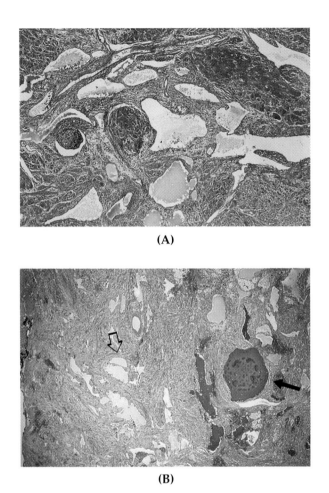

(A)

(B)

Figure 5.33
Deep lymphatic malformations. **(A)** Cavernous lymphatic channels with mural lymphoid tissue and thin walls in a deep soft tissue lesion excised from the face of a 5-year-old child. **(B)** This intramuscular malformation contains venous (solid arrow) and lymphatic (open arrow) components.

gins, therefore, should be examined carefully because recurrences are common after incomplete excision (Coffin, 1997).

Cystic hygromas are a morphological variant of lymphatic malformations that occur most commonly in the cervical area. Histologically, they differ from the so-called cavernous lymphangiomas only in greater dilation of lymphatic spaces (Fig. 5.34) (Emery et al., 1984). The predilection of these variants for the neck may reflect lower resistance to cystic expansion in the characteristically loose connective tissue of

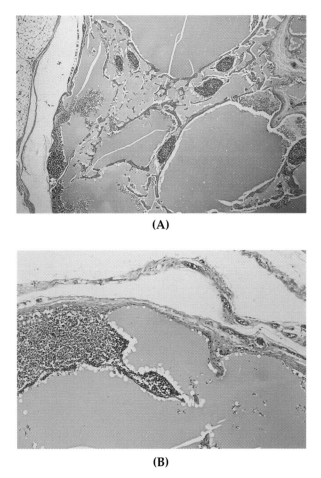

(A)

(B)

Figure 5.34
Cystic hygroma. **(A)** Low-power view of a neck lesion, demonstrating confluent, grossly enlarged lymphatic spaces with mural lymphoid tissue. **(B)** High-power view of same lesion as in A. Note valve-like structure and mural lymphoid tissue protruding into lumen filled with proteinaceous fluid containing lymphocytes.

that area. Lymphatic malformations in the neck and adjacent structures sometimes resemble branchial cleft and thymic cysts, which also may have attenuated, flattened linings, lumina filled with proteinaceous fluid, and a surrounding inflammatory response. Cholesterol granuloma formation is common in the connective tissue surrounding branchial cleft and thymic cysts (Fig. 5.35), however, but is not seen in lymphatic malformations.

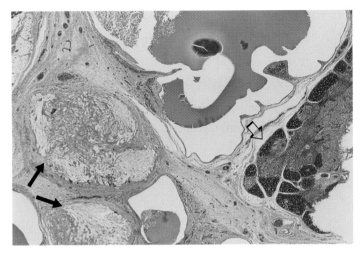

Figure 5.35
Thymic cyst. Multiloculated thymic cyst excised from anterolateral neck of a 5-year-old child. Note thymic tissue (open arrow), cholesterol granulomas (solid arrows), and attenuated cyst lining. Cholesterol granulomas are a distinguishing feature, as they are not seen in lymphatic malformations.

Lymphatic malformations located in subcutaneous tissues generally have normal overlying skin. However, many lymphatic malformations in the dermis are associated with deeper subcutaneous portions composed of larger vessels. This association has suggested that the ectatic dermal lymphatics of "lymphangioma simplex" represent outpocketings from larger subcutaneous lymphatics (Whimster, 1976). This pathogenic mechanism would explain the frequent recurrence of these dermal lesions after skin grafting.

Lymphatic malformations commonly include a significant venous component (Fig. 5.33B). This is perhaps best explained by the close relationship between the lymphatic and venous systems during embryonic development. Such lymphatic–venous lesions are discussed below in the section on mixed low-flow malformations. It has been suggested that these mixed lymphatic and venous malformations are more locally aggressive than pure lymphatic malformations (Szilagyi et al., 1976; Grabb et al., 1980). Lymphatic malformations, particularly those with a venous component, occasionally have been noted to regress in size. This apparent "involution" is probably a deflation of engorged vessels rather than a true involution by cell death and vessel drop-out as is seen in hemangiomas, and it may result from spontaneous formation of microscopic venous–lymphatic communications.

5.3.2. Venous Malformations

Venous malformations are a diverse group of lesions characterized by abnormal collections of veins. The component veins vary with regard to lumenal size, wall thickness, and degree of abnormality in mural architecture both between and within individual lesions. Venous malformations may be superficial or deep, diffuse or localized. Some pathologists have used the term *cavernous hemangioma* to refer to venous malformations that demonstrate irregularly shaped, very thin-walled vessels of large lumenal diameter, while reserving the term *venous malformation* for lesions composed of thicker walled veins with obviously disordered smooth muscle architecture. Such usage, however, perpetuates misuse of the term *hemangioma*, as previously discussed, and obscures the fact that both of these histological appearances do, in fact, represent malformations of veins. Port-wine stains are a clinically distinct subtype of venous malformation composed of dilated venule-like channels within the dermis and are discussed separately.

Multiple venous malformations characterize the dysmorphic syndrome known by the unforgettable name of *blue rubber bleb nevus syndrome*. This clinical entity was first described in 1958 by Bean as an association between multiple venous malformations of the skin and of the gastrointestinal tract complicated by gastrointestional bleeding; it is now thought to be a diffuse venous anomaly, possibly encompassing multiple entities, which can involve multiple viscera (for review, see Young, 1988b). Venous malformations have also been described in Turner's syndrome, both in the gastrointestinal tract (Burns et al., 1991) and on the dorsum of the feet (Weiss, 1988).

Venous malformations, like all true vascular malformations, have no demonstrable endothelial or pericytic mitotic activity. Instead, they demonstrate, in lesions of all ages, an abnormally high density of veins lined by a flattened, mature-appearing endothelium. The venous nature of these malformations is implied, quite simply, by the presence of a variable amount of smooth muscle in the vessel walls (usually scant relative to luminal diameter), the absence of an internal elastic membrane, and component lumina filled erythrocytes (Figs. 5.36–5.39). Vessels of capillary or venular proportions may also be dispersed within the lesion, sometimes forming loose conglomerates that lack the

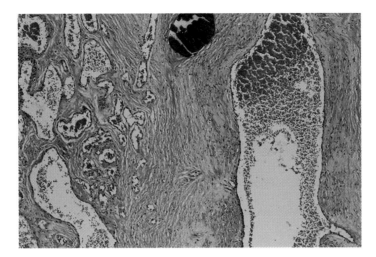

Figure 5.36
Venous malformation. The constituent vascular elements are thin-walled veins of various diameters. Note calcified thrombus.

well-defined lobularity of hemangiomas. Deep lesions that also involve the overlying dermis often show a higher proportion of smaller vessels in the dermal component than in deeper areas. The walls of larger lesional vessels may be irregularly thickened by adventitial fibrosis, producing marked and abrupt variations in wall thickness around the

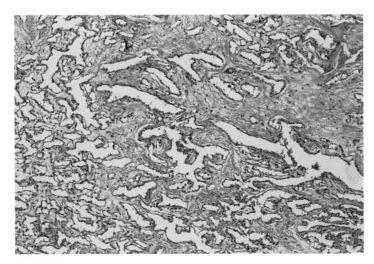

Figure 5.37
Venous malformation. An anastomosing network of thin-walled veins characterizes this lesion excised from the periorbital soft tissue of a 14-day-old infant. Note the flat, nonproliferative endothelium.

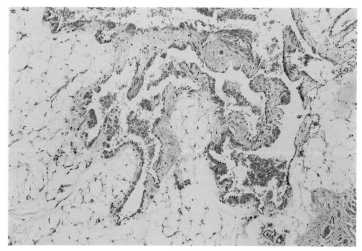

Figure 5.38
Venous malformation. Partially collapsed, "cavernous" veins with flat endothelia and irregular, but generally thin walls are found within this facial subcutaneous lesion from a 2-year-old child. The lumina of these vessels are disproportionately large relative to the amount of mural smooth muscle present.

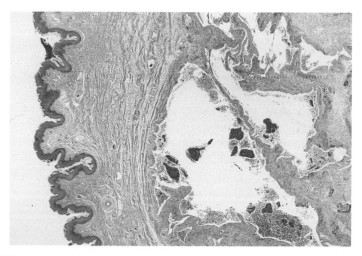

Figure 5.39
Venous malformation. This subcutaneous lesion, excised from the scrotum of a 14-year-old boy, enlarged during puberty. There is a central cluster of abnormal, blood-filled veins.

vessel circumference. Venous malformations excised from some patients, particularly those with multiple lesions, may show extreme disorganization of component smooth muscle fibers, which focally stream out haphazardly into the surrounding connective tissue (Fig. 5.40). Pronounced thickening of venous walls suggests the possibility of abnormal arteriovenous shunting within the lesion and prompts clinical and radiological correlation.

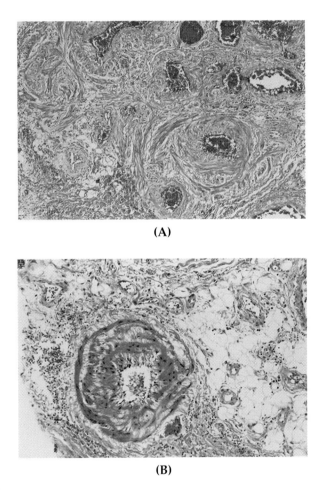

(A)

(B)

Figure 5.40
Venous malformation. Low-power **(A)** and high-power **(B)** views show markedly disorganized vascular smooth muscle in this lower extremity lesion from a 2-year-old child with extensive bilateral malformations.

Venous malformations are generally not as well circumscribed as hemangiomas. It is important to examine carefully the margins of resection for abnormal vessels because these lesions, like other vascular malformations, tend to recur by progressive ectasia of residual vessels following incomplete resection. The connective tissue stroma between lesional vessels is often fibrotic and may show focal chronic inflammation. This inflammation is usually minimal, but is more prominent in intramuscular venous malformations and in those with a significant lymphatic component (see section below on mixed malformations).

Luminal thrombi, in various stages of organization, are common in venous malformations (Fig. 5.41), presumably as a result of stasis in these low-flow lesions, and may show dystrophic calcification. Organizing thrombi undergoing recanalization occasionally demonstrate a reactive intravascular papillary endothelial hyperplasia that is restricted to the lumina of involved vessels (Fig. 5.42–5.45). This phenomenon, probably equivalent to that first described in 1923 for hemorrhoidal veins (Masson, 1923), has been referred to previously

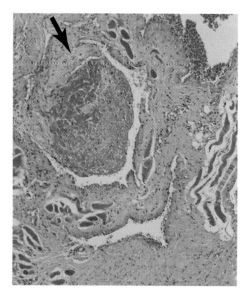

Figure 5.41
Venous malformation. Thrombosis is common in venous malformations. Note organizing thrombus attached to vessel wall (arrow).

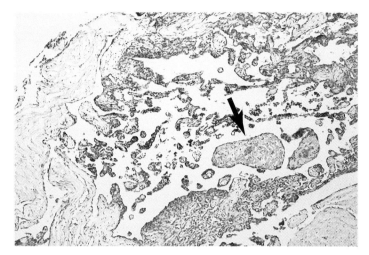

Figure 5.42
Venous malformation, with intravascular papillary endothelial hyperplasia. Connective tissue papillae, lined by endothelial cells, have formed within the lumen of a large vessel in a venous malformation. This benign proliferative process is thought to result from organization and endothelialization of thrombus material. Note the core of residual organizing thrombus in one of the papillae (arrow).

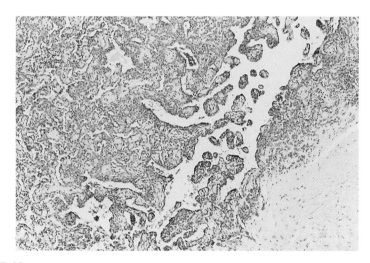

Figure 5.43
Venous malformation, with intravascular papillary endothelial hyperplasia. Hyperplastic endothelial cells lining the papillary cores are plump, but do not show atypia. Mitotic activity is not evident in this example.

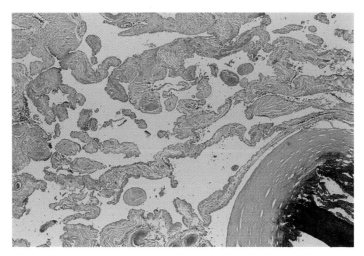

Figure 5.44
Venous malformation, with intravascular papillary endothelial hyperplasia. This mature lesion shows fibrotic papillary cores and flattened, inactive endothelia. Note adjacent calcified thrombus.

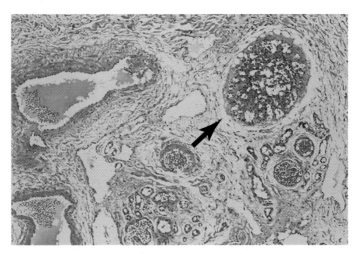

Figure 5.45
Venous malformation, with intravascular papillary endothelial hyperplasia. Circumscribed foci of intravascular papillary endothelial hyperplasia, such as the one shown here in a venous malformation (arrow), may be misinterpreted as a lobular hemangioma (see text).

as *Masson's vegetant intravascular hemangioendothelioma* or *Masson's pseudoangiosarcoma* (Clearkin and Enzinger, 1976; Hashimoto et al., 1983). The endothelial cells in this reactive process may be plump, but do not demonstrate mitotic figures or significant nuclear pleomorphism; they line up around hyalinized eosinophilic stalks or cores in a papillary or sinusoidal pattern, forming a sometimes complex network of anastomosing vascular channels. Although this type of lesion, due to its etiological association with organizing thrombi, has a predilection for pre-existing venous malformations, it can also present as a mass within an apparently normal vessel of skin, subcutaneous, or deep soft tissue, most commonly in adults; rarely, it occurs extravascularly, possibly within organizing hematomas (Pins et al., 1993). It is important in any context to differentiate this benign process from malignant processes such as angiosarcoma, Kaposi's sarcoma, and endovascular papillary angioendothelioma. In a child it is especially important to recognize and distinguish this process, which may be multifocal in venous malformations, from lobules of involuting hemangioma (Fig. 5.45). Enclosure by a vessel wall is usually obvious at low power, and closer inspection may reveal continuity between the eosinophilic matrix of the papillary cores and recognizable thrombus at the periphery of the lesion.

Glomangiomas are vascular tumors occurring in childhood that strongly resemble venous malformations, augmented by a few layers or nests of round to oval glomus cells with eosinophilic cytoplasm around the component thin-walled, gaping veins (Fig. 5.46A). As in venous malformations, organizing thrombi are common (Fig. 5.46B). These tumors of childhood can be congenital and are clinically as well as histologically distinct from adult-type, typically subungual glomus tumors. They are often multiple, with evidence of autosomal dominant inheritance, and are painless; congenital forms are plaque-like in appearance and, like venous malformations, enlarge slowly with body growth (Schirren et al., 1993; Kohout and Stout, 1961; Wood and Dimmick, 1977; Landthaler et al., 1990; Troschke et al., 1993) (For review, see Glick et al., 1995). The identity of the glomus cells is confirmed by immunohistochemical studies that show immunoreactivity for SMC-actin and vimentin, but not for vWf and desmin.

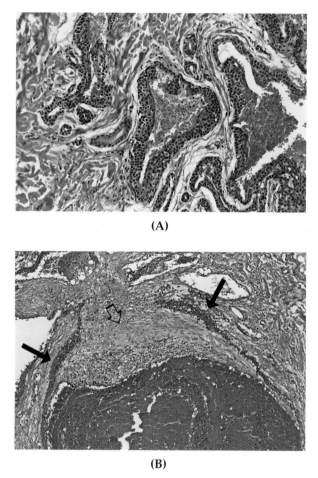

Figure 5.46
Glomangioma. **(A)** This distinct variant of a venous malformation consists of cavernous veins surrounded by one or more layers of glomus cells. **(B)** Glomangiomas, like venous malformations, are prone to thrombosis. Note perivascular rim of glomus cells (solid arrows) disrupted by an organizing thrombus (open arrow).

5.3.3. Port-Wine Stains (Cutaneous Venulocapillary Malformations)

Port-wine stains are congenital skin lesions ("birthmarks") characterized histologically by ectatic vessels of capillary or venular size within the papillary and superficial reticular dermis. They are nonproliferative lesions that do not involute and are therefore best categorized as vascular malformations. They do, however, change in both histological and clinical appearance with age, becoming more nodular and raised as vessel ectasia progresses to include the deeper dermal and superficial subcutaneous tissues. Port-wine stains have traditionally been included under the term *nevus flammeus*, along with the very common *nevus flammeus neonatorum*, known colloquially as *salmon patch, stork's bite*, or *angel's kiss*. These two entities differ, however, in a number of important aspects. Salmon patches are extremely common in newborns, occur as pink macules, most commonly on the nape of the neck, eyelids, and glabella, and often fade or disappear completely with time, leaving no histological residuum. The color of these "macular stains" transiently deepens with exercise or emotional upset, and these lesions may represent focal areas of physiological vascular dysfunction, such as poor precapillary sphincter control, rather than true malformations. Port-wine stains, on the other hand, are relatively rare macular lesions (occurring in about 0.3% of newborns [Esterly, 1987]), generally present on the skin of the head and neck and often within the distribution of the trigeminal nerve. They persist throughout life, grow with the child, and become raised and darker with age. In unusual cases, they are associated with ipsilateral leptomeningeal and choroid venous malformations (Sturge-Weber syndrome). Port-wine stains may sometimes directly overlie deeper malformations of large vessel type in other regions of the body, as in association with Klippel-Trenaunay syndrome and Parkes-Weber syndrome (Lindenauer, 1965).

Biopsies of port-wine stains from infants and young children may not reveal the characteristic vessel ectasia, which does not begin to become prominent until about 10 years of age (Finley et al., 1984). From that time onward, capillary- and venule-sized vessels in the upper dermis become progressively dilated and filled with erythrocytes (Figs. 5.47, 5.48). These vessels are lined by flat, inactive-appearing endothelia, without evidence of mitotic activity. While there is some controversy as to whether dermal vessels are actually increased in number in port-wine stains compared with normal skin (Barsky et al., 1980; Smoller and Rosen, 1986), there is no evidence that the number of vessels in these lesions changes with aging. Ectasia begins superficially within the lesions and progressively involves deeper vessels, eventually extending into the reticular dermis and focally into the subcutaneous tissue. Immunoreactions for vWf, collagenous basement membrane proteins, and fibronectin have not shown differences between

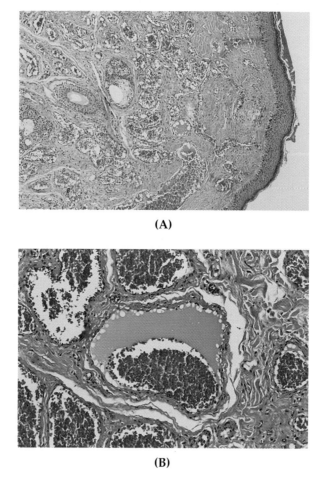

(A)

(B)

Figure 5.47
Port-wine stain. Low-power **(A)** and high-power **(B)** views of a port-wine stain from a 19-year-old person show ectatic vessels of capillary to venular size in the upper dermis. Note thin vessel walls and flat endothelia.

port-wine stains and normal skin (Finley et al., 1982). However, immunoreaction for S-100, a protein found in Schwann cells, reveals a significant decrease in perivascular nerve density in port-wine stains, suggesting that inadequate innervation may be responsible for the progressive vascular dilation that is characteristic of these lesions (Smoller and Rosen, 1986).

Salmon patches are usually not biopsied due to their ease of clinical recognition and their benign and usually transient nature. Biopsied lesions show dilated capillaries of otherwise normal morphology in the papillary dermis (Calonje and Wilson-Jones, 1997).

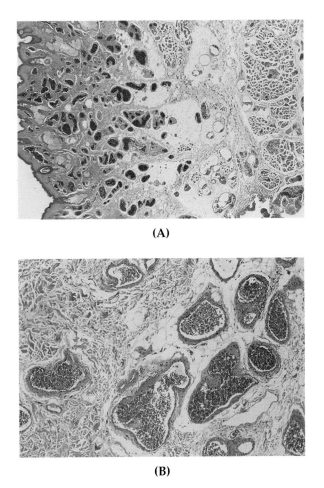

(A)

(B)

Figure 5.48
Port-wine stain. Low-power **(A)** and high-power **(B)** views of a port-wine stain from the face of a 60-year-old adult that had become raised and nodular with age. Blood-filled vessels fill the papillary and reticular dermis and extend into the sub-cutaneous adipose tissue. Note mild thickening of vessel walls (B).

5.3.4. Arteriovenous Malformations

Arteriovenous malformations occurring in children are usually evident at birth and are associated with a variable, but often clinically significant degree of arteriovenous shunting. They are deep lesions that are encountered most commonly in the head, neck, or extremities. Lesions located close to the skin surface may produce a palpable thrill or pulsation and elevation of skin temperature. Many of these lesions have associated venulocapillary malformations resembling port-wine stains in the overlying dermis (Gomes and Bernatz, 1970; Szilagyi et al., 1976)—hence the adage "pulsatile birthmark." Other synonyms which have been applied to these lesions include "arteriovenous aneurysm," "circoid aneurysm," "racemose hemangioma," "arteriovenous hemangioma," and "pulsating angioma." Because these lesions do not arise from proliferative processes, but are instead developmental errors of morphogenesis without histological evidence of cellular hyperplasia, we prefer the term *arteriovenous malformation*.

The histological appearance of arteriovenous malformations varies widely from one area to another, and the actual arteriovenous shunts, which are tiny and numerous, are difficult to demonstrate without laborious sectioning or special techniques. For these reasons, it is imperative to consider the clinical and radiological findings, as well as histological appearance, when diagnosing these lesions. Arteriovenous malformations "grow" with the child by progressive enlargement of feeding and draining vessels, as well as "recruitment" of collateral vasculature. The histological picture, particularly of older lesions, may be dominated by secondary change, obscuring the originating nidus of abnormal arteriovenous communications. Successful surgical excision of these lesions is dependent on removal of the entire nidus of arteriovenous shunting, thus precluding recruitment of new arterial collaterals to refill the residual nidus and continue the process. Unfortunately, the margins of the nidus are difficult to define histologically, and the radiologist is of more assistance to the surgeon than is the pathologist in directing the extent of resection. The pathologist is essential, however, in confirming the diagnosis of arteriovenous malformation. We have seen a number of deeply located, proliferative hemangiomas that were mistaken clinically for localized arteriovenous malformations, perhaps because of significant arteriovenous shunting occurring through the newly formed, dilating capillary beds of the lesion.

Arteriovenous malformations excised from very young children may show only increased numbers of capillaries, venules, and arterioles, loosely grouped within a fibrous or fibromyxomatous background, with few larger caliber vessels. These vessels may be dilated and filled with erythrocytes, and the stroma may show hemosiderin deposition. The vessel endothelia are flat and mature appearing, without mitotic figures. Some lesions demonstrate an apparent margin where the bed of small vessels merges with surrounding, normally vascularized connective tissue. Lesions excised from older children and adults show a greater prominence of larger vessels, with fewer capillaries evident in most sections. Enlarged veins and sometimes tortuous arteries, medium sized or larger, may be seen in close proximity to one another (Figs. 5.49, 5.50). The veins show adventitial fibrosis, medial hypertrophy, and intimal hyperplasia, and an elastic stain may be necessary to distinguish these vessels from arterial elements (Fig. 5.50). The asymmetical "arterialization" of component veins that occurs in arteriovenous malformations reflects increased venous pressure due to arterial shunting and is more pronounced than the irregular fibrous wall thickening sometimes seen in venous malformations. Dilated lymphatics may be present, but are usually not prominent. The overlying skin may be involved by the lesion, primarily in the form of a venulocapillary component similar to that of port-wine stains, and may also show secondary changes, including necrosis due to arterial steal and kaposiform changes known as *kaposiform angiodermatitis* or *acroangiodermatitis* (Bluefarb and Adams, 1967; Earhart et al., 1974; Rusin and Harrell, 1976; Strutton and Weedon, 1987). These latter changes represent a reaction to the underlying shunt and consist of a proliferation of thickwalled capillaries with rounded lumina and plump endothelial cells in the superficial dermis, with adjacent accumulations of fibroblasts and hemosiderin.

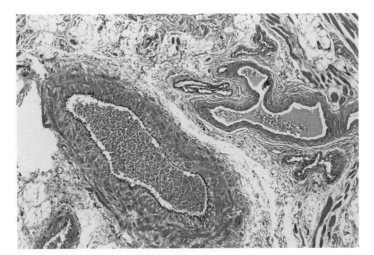

Figure **5.49**
Arteriovenous malformation. Arteries and thick-walled, "arterialized" veins may be difficult to distinguish in H&E-stained sections. This high-flow malformation was excised from the lip of a 6-year-old child.

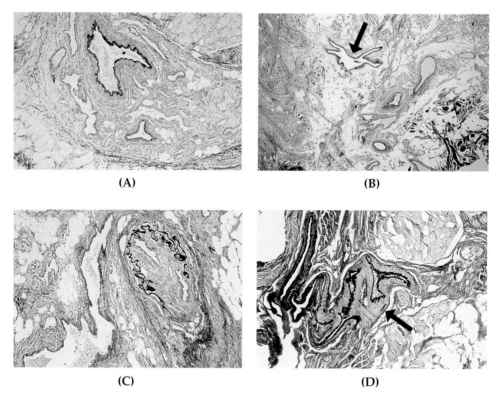

Figure 5.50
Arteriovenous malformation. An elastic stain helps differentiate arteries from "arterialized" veins by highlighting the internal elastic lamina of the arteries. **(A)** Note branched artery surrounded by a dense grouping of smaller, irregularly shaped veins. **(B)** Note complex, branched vein (solid arrow) and smaller artery with well-organized internal elastic lamina. **(C)** Note atypical vessel with fragmented elastic lamina, adjacent to a large vein. **(D)** Tortuosity of arterial feeder vessels, as demonstrated here (arrow), represents an effect, rather than a cause, of high-flow shunting in arteriovenous malformations.

Direct continuities between arteries and veins in arteriovenous malformations are difficult to identify in routine sections, but these have been documented in careful studies (Lawton et al., 1957). Individual communications are microscopic, but may number in the millions in large lesions. In contrast, acquired arteriovenous fistulae, resulting from trauma or surgical intervention, are generally single. They are also generally accompanied by histological evidence of the traumatic event.

5.3.5. Mixed, Low-Flow Vascular Malformations

Mixed vascular malformations that exhibit low-flow characteristics consist of combinations of venous, lymphatic, and capillary components in variable proportions. Vessels of capillary and venular size are so commonly a component of venous malformations that they usually need not be specifically mentioned in the diagnosis. Exceptions to this are those deep, large vessel lesions that include an overlying dermal component composed of a predominantly small vessel malformation. In this case, a diagnosis such as "subcutaneous venous–lymphatic malformation with dermal capillary component" will explain the surface appearance of the lesion to the surgeon, as well as define the underlying deep vessel component that will determine the clinical behavior of such a lesion. Designation of a mixed malformation as "venous–lymphatic" versus "lymphatic–venous," according to the relative proportions of the two components in the main lesion, is also appropriate. The historical term *lymphangiohemangioma*, however, is not appropriate. These lesions are clearly malformations, without evidence of cellular hyperplasia in either component.

The histology of mixed lesion, low-flow malformations is essentially a variable merging of the histological appearances of the pure lesions, depending on the relative proportions of the constituent elements, and does not require great elaboration. Both blood- and lymph-filled vessels are seen (Fig. 5.33B), and spontaneously occurring lymphatic–venous fistulas have been postulated to explain the occasional observation of mixing of erythrocytes and lymph within some vessels, beyond that expected of artifact. Greater prominence of lymphatic components correlates with increased stromal chronic inflammation. Lymphatic–venous malformations do not usually involve the skin, but can, like pure lymphatic malformations, provoke dilation of communicating mucosal or cutaneous lymphatics. A number of dysmorphic syndromes have been associated with complex combinations of lymphatic, venous, and capillary malformations, including Maffucci's syndrome and Klippel-Trenaunay syndrome (Burns et al., 1991).

5.3.6. Pathogenesis of Vascular Malformations

Vascular malformations originate from errors in morphogenesis of the embryonic vasculature. It is likely that these malformations further mold themselves, and perhaps the tissues that surround them, through the altered hemodynamic forces arising within the lesion. These effects are most evident with arteriovenous malformations. The embryonic development of the vascular system is influenced by many factors, including genetics, biochemical and chemical microenvironments, blood flow, mechanical stresses, endothelial–pericytic and endothelial–smooth muscle interactions, and the developing autonomic nervous system. Experimental administration of toxins, including known teratogens, has not succeeded in producing vascular malformations in animal fetuses, suggesting that simple abnormalities in local microenvironment are not responsible (Young, 1988a). The known association of vascular malformations with a number of genetic syndromes has offered the promise of identification, through linkage analysis, of specific protein abnormalities involved in vascular morphogenesis. Exciting progress has been made recently in this area by Vikkula and colleagues (Vikkula et al., 1996; Gallione et al., 1995; Boon et al., 1994). They mapped dominantly inherited, multiple mucocutaneous venous malformations (clinically synonomous with *blue rubber bleb nevus syndrome*) in two unrelated families to a missense mutation in the kinase domain of the receptor tyrosine kinase TIE2, localized to chromosome 9p21 (Dumont et al., 1994). Furthermore, they demonstrated that this mutation results in increased activity of TIE2 kinase and suggested that this activation is responsible for the poor development of smooth muscle seen in the component vessel walls of venous malformations occurring in these families, perhaps by influencing PDGF/TGF-β–mediated recruitment of smooth muscle cells. While these key findings may not bear directly on the pathogenesis of other caregories of vascular malformations, they significantly advance the understanding of molecular mechanisms of vascular morphogenesis and suggest exciting new possibilities for research.

The potential role of the nervous system in the pathogenesis of vascular malformations also deserves further study. The aforementioned observation by Smoller and Rosen (1986) that perivascular nerves are decreased in number in port-wine stains is intriguing and suggests that similar analyses should be applied to other types of vascular malformations. Robinson and coworkers (1994) in a recent similar study explored the immunoreactivity of intramuscular "hemangiomas," including two lesions from a single patient with the rare Maffucci's syndrome, for S-100 and a battery of neuropeptides. They found an increased number of nerve fibers in solitary "hemangiomas" (from otherwise normal patients), as well as in those occurring in association with Maffucci's syndrome, compared with normal muscle. Furthermore, this increase in nerve density extended past the margins of the lesions in the patient with Maffucci's syndrome, but not in those with sporadic hemangiomas. This otherwise careful study is unfortunately limited by the imprecision of histopathological classification that plagues this area of research. Skeletal muscle "hemangiomas" are usually vascular malformations, although true proliferative hemangiomas do occur in muscle, and the vascular lesions of Maffucci's syndrome are venous or lymphatic–venous malformations rather than hemangiomas. Whether these emerging neural abnormalities represent cause or effect in the pathogenesis of vascular malformations remains problematic.

5.4. RARE VASCULAR LESIONS OF INFANCY ASSOCIATED WITH KASABACH-MERRITT SYNDROME

Kasabach-Merritt syndrome is a potentially devastating bleeding diathesis defined as thrombocytopenic purpura complicating a large hemangioma or "hemangioma-like" tumor (Kasabach and Merritt, 1940). This syndrome is marked by profound thrombocytopenia resulting from platelet-trapping within the tumor, sometimes compounded by secondary consumption of fibrinogen and coagulation factors. It is distinct from the chronic consumptive coagulopathy that occurs in venous malformations due to blood stasis in which platelet counts are normal or only modestly decreased.

Three rare, histologically distinct lesions, which may be congenital or may present in infancy, have been described in association with Kasabach-Merritt syndrome. These lesions, which do not fall squarely under the headings of hemangioma or vascular malformation, are (1) the recently described infantile kaposiform hemangioendothelioma; (2) tufted angioma (angioblastoma); and (3) infantile hemangiopericytoma. The first two entities are sometimes found in association with lymphatic malformations and appear to be the major causes of the coagulopathy that characterizes Kasabach-Merritt syndrome (Enjolras et al., 1997).

5.4.1. Infantile Kaposiform Hemangioendothelioma

This rare, locally aggressive vascular tumor has only recently been recognized as a distinct histological entity (Tsang and Chan, 1991; Zukerberg et al., 1993), although similar lesions have previously been reported under different names, including the *capillary hemangioma* originally described by Kasabach and Merritt in association with the syndrome that now carries their names (Kasabach and Merritt, 1940; Niedt et al., 1989; Al-Rashid, 1971; Pearl and Mathews, 1979; Weinblatt et al., 1984; Lai et al., 1991). These tumors are classified as borderline malignant (hemangioendotheliomas) due to their locally aggressive behavior, but distant metastasis is thought not to occur. The small number of reported cases suggests equal frequencies in both sexes and includes both superficial and deep examples that have arisen in the head and neck, extremities, retroperitoneum, mediastinum, and chest wall. Some patients with these tumors do not develop Kasabach-Merritt syndrome, probably reflecting site and extent of disease, and it seems likely that many such tumors arising in patients without the full syndrome have gone unreported, probably classified simply as unusual hemangiomas (Zukerberg et al., 1993). The biological behavior of these lesions, and their potential for spontaneous regression, are incompletely understood. Some have responded well to therapy with steroids or interferon-α2a (Ezekowitz et al., 1992).

Reported examples of infantile kaposiform hemangioendothelioma typically have been large, with an infiltrative, nodular growth pattern. The tumor nodules are composed primarily of fascicles of moderately plump spindle cells with eosinophilic-to-clear cytoplasm and bland nuclei demonstrating variable, but generally low degrees of mitotic activity (Figs. 5.51–5.53). The spindled areas may be interrupted by lobules of small, irregularly dilated capillaries containing erythrocytes and lined by flat endothelia. Abnormal groupings of dilated lymphatic channels also occur commonly within or around the tumor, suggesting the possibility of a co-existent lymphatic malformation (Fig. 5.51) (Bircher et al., 1994; Boukobza et al., 1996). Cavernous blood-filled spaces, typical of spindle-cell hemangioendothelioma, have not been described for these lesions. The spindle cells contain small amounts of hemosiderin and focally form elongated, slit-like lumina containing red blood cells, reminiscent of Kaposi's sarcoma. These lumina also may contain microthrombi. Some examples also contain nests of rounded,

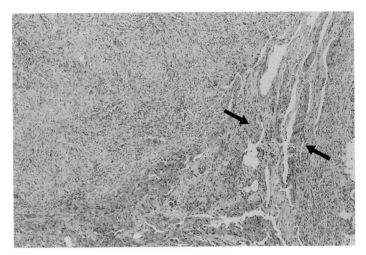

Figure 5.51
Infantile kaposiform hemangioendothelioma. Low-power view shows large, ill-defined nodules of plump spindle cells forming tiny lumina containing erythrocytes. Note abnormal lymphatic spaces (between arrows) within the tumor, coursing between nodules.

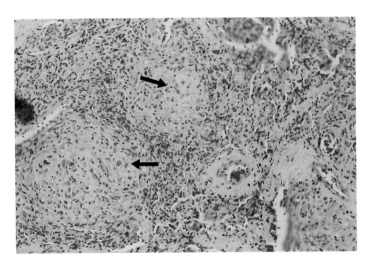

Figure 5.52
Infantile kaposiform hemangioendothelioma. High-power view shows occasional differentiated vessels within the tumor (lower right) and two nodules containing a few epithelioid cells (arrows).

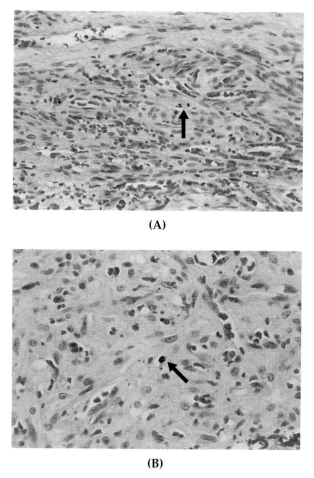

(A)

(B)

Figure 5.53
Infantile kaposiform hemangioendothelioma. Mid-power **(A)** and high-power **(B)** views of the lesion shown in Figure 5.52 show slit-like lumina formed by relatively bland-appearing spindle cells. Scattered mitotic figures (arrows) are present in this example excised from the parotid region of a 4-month-old infant with Kasabach-Merritt syndrome.

plump endothelial cells similar to those seen in epithelioid heman-gioendothelioma (Zukerberg et al., 1993), but this feature is not univer-sal (Enjolras et al., 1997). Lesions involving the skin are apt to be surrounded by dense fibrosis, but lack the plasma cell-rich chronic inflammatory infiltrate characteristic of Kaposi's sarcoma. Immuno-histochemical studies reveal strong expression of CD34 and weak or absent expression of vWf by the spindle cell population (Zukerberg et al., 1993; Tsang and Chan, 1991). Only scattered pericytes are demonstrable within the fascicles of spindle cells by electron microscopy (Zukerberg et al., 1993) and actin immunoreactivity (Zukerberg et al., 1993; Tsang and Chan, 1991). Well-formed vascular structures within the lesion predictably show immunolabeling patterns characteristic of mature endothelium, as well as cuffs of pericytes expressing actin. Electron microscopy of the spindled endothelial cells demonstrates interlocking processes and junctions, but poorly formed basement membranes (Zukerberg et al., 1993).

5.4.2. Tufted Angioma (Angioblastoma of Nakagawa)

These benign vascular tumors are classified appropriately as hemangiomas, but they differ from classic juvenile hemangiomas in several important clinical and pathological respects. First described in 1949 by Nakagawa and later by others (Miki and Matsumoto, 1962; MacMillan and Champion, 1971; Kumakiri et al., 1983; Jones and Orkin, 1989), these clinically and histologically distinctive lesions have been called many things other than angioblastoma, including "progressive capillary hemangioma," "hypertrophic hemangioma," "tufted angioma," and "tufted hemangioma." The term *tufted angioma* is currently favored, often qualified as *acquired tufted angioma* to reflect the usual appearance of these lesions after the first year of life and sometimes in adulthood. However, 15% of these lesions are present at birth (Kumakiri et al., 1983; Jones and Orkin, 1989). An autosomal dominant pattern of inheritance has been reported for one family (Heagerty et al., 1992). The tumors appear as erythematous plaques or macules located on the upper portion of the body that grow slowly for several years before stabilizing. Regression has occurred in rare instances (Lam et al., 1994). Tufted angiomas, like infantile kaposiform hemangioendotheliomas, are equally prevalent in both sexes and thus do not show the female predominance of juvenile hemangiomas.

The characteristic pattern of this lesion is a "cannonball" distribution of rounded nodules or tufts of capillary-sized vessels in the middermis, sometimes extending into the subcutaneous fat (Jones and Orkin, 1989). The tufts are irregularly distributed throughout the dermis, and may be surrounded by dense fibrosis. The endothelial cells in each tuft are tightly packed, with inconspicuous lumina, except at the periphery of the tuft where a few larger, blood-filled vessels may be present, compressed into crescents by the dense, proliferating central "ball" of capillaries. Mitotic figures are not seen in typical lesions, reflecting their slow growth. The endothelial cells are immunoreactive for *Ulex europaeus* lectin I. However, those in the most solid areas may not express factor vWF, a result also seen in early juvenile hemangiomas, probably as a consequence of cellular immaturity. Both immunohistochemical (Padilla et al., 1987) and electron microscopic studies (Kimura, 1981) have documented a significant population of pericytes intermixed with the endothelial cells. The latter studies have also shown lamellar crystalline inclusions within endothelial cytoplasm, similar to those seen in juvenile hemangiomas and in fetal endothelum (Kumakiri et al., 1983). Interestingly, all reported examples of this rare tumor occurring in association with Kasabach-Merritt syndrome have arisen in infants and have included abnormal lymphatic spaces intermingling with or adjacent to the angiomatous lesion (Enjolras et al., 1997). This association mimics the previously mentioned co-existence of lymphatic malformations and infantile kaposiform hemangioendotheliomas, now reported by two independent groups (Enjolras et al., 1997; Zukerberg et al., 1993). Tufted angiomas and infantile kaposiform hemangioendothelioma do share some histological features, suggesting that they may be variants within the same pathological group rather than distinct entities. Enjolras and coworkers (1997) report a vascular lesion associated with Kasabach-Merritt syndrome in which features consistent with tufted angioma were combined focally with infiltrating spindle cells reminiscent of infantile kaposiform hemangioendothelioma.

5.4.3. Infantile Hemangiopericytoma

Infantile hemangiopericytomas differ from those found in adults in a number of fundamental clinical and pathological aspects. They usually present at birth or shortly thereafter, are more common in males, and are located predominantly in the subcutaneous tissue and dermis of the head and neck (Kauffman and Stout, 1960; Enzinger and Smith, 1976; Bailey et al., 1993). Multifocality has been reported (Mentzel et al., 1994; Seibert et al., 1978), as have both local recurrence following resection (Atkinson et al., 1984; Alpers et al., 1984) and spontaneous regression (Chen et al., 1986). Rare cases with apparent distant spread have also occurred (Morgan and Evbuomwan, 1983), but may reflect mutilfocality rather than true metastasis. Infantile hemangiopericytomas are generally treated conservatively due to their benign behavior, but the clinical course can be complicated by hemorrhage, and these tumors may threaten adjacent structures by their rapid growth. Two cases have been reported in association with possible Kasabach-Merritt syndrome (Resnick et al., 1993; Chung et al., 1995). Infantile hemangiopericytomas are currently thought of as part of a spectrum of infantile myofibroblastic lesions that also includes infantile myofibromatosis, within which the specific histological pattern varies with degree of lesion maturity (Coffin and Dehner, 1991; Mentzel et al., 1994).

Infantile hemangiopericytomas are multilobular lesions demonstrating intravascular or perivascular extension beyond the main tumor mass (Coffin, 1997). Microscopically, the lobules of tumor show fasicles of plump spindle cells with eosinophilic cytoplasm, alternating with more cellular areas containing less mature, rounded cells with hyperchromatic nuclei (Fig. 5.54). The latter cells indent vascular lumina within the mass, producing the characteristic hemangiopericytoma-type pattern (Fig. 5.55), and mitotic activity may be brisk. The spindle cells immunoreact with antibodies to SMC-actin, whereas the less mature cells may not (Mentzel et al., 1994; Ordonez et al., 1993).

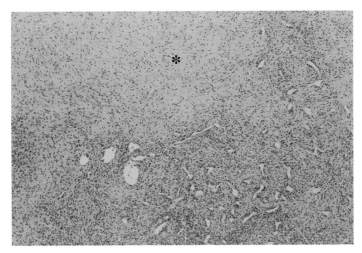

Figure **5.54**
Infantile hemangiopericytoma. Areas of plump spindle cells with eosinophilic cytoplasm (asterick) alternate with more-cellular areas containing small vessels surrounded by small, rounded cells.

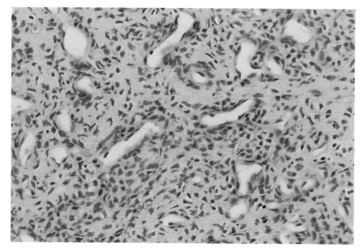

Figure 5.55
Infantile hemangiopericytoma. Vascular areas of the tumor show a characteristic
hemangiopericytoma-type pattern consisting of small vessels, lined by endothelial
cells and surrounded by immature cells with hyperchromatic nuclei. These tumors
are thought to be myofibroblastic in origin (see text).

5.5. CONCLUSION

In this chapter we attempt to delineate clearly the anatomical and
pathological features of the various vascular tumors and malformations that occur in children. Our aim is to provide clear pathological description as a basis for clinical classification and treatment of these lesions and to facilitate research into understanding their behavior.

REFERENCES

Albelda, S.M., Smith, C.W., and Ward, P.A.: Adhesion molecules and inflammatory injury. FASEB J. 8:504–512, 1994.

Alessi, E., Bertani E., and Sala, F.: Acquired tufted hemangioma. Am. J. Dermatopathol. 8:426–429, 1986.

Allen, P.W., and Enzinger, F.M.: Hemangiomas of skeletal muscle: An analysis
of 89 cases. Cancer 29:8, 1972.

Alles, J.U.: Immunzytochemische charakterisieruug vascularer tumoren.
Habilitation sschrift des Fachbereichs Humanmedizin der Justus-Liebig-
Universitat GieBen, 1987.

Alpers, C.E., Rosenau, W., Finkbeiner, W.E., deLorimer, A.A., Kronish, D.: Congenital (infantile) hemangiopericytoma of the tongue and sublingual region. Am. J. Clin. Pathol. 81:377–382, 1984.

Al-Rashid, R.A.: Cyclophosphamide and radiation therapy in the treatment of disseminated intravascular clotting. Cancer 27:364–368, 1971.

Amir, J., Metzker, A., Krikler, R., and Reisner, S.H.: Strawberry hemangioma in preterm infants. Pediatr. Dermatol. 3:131–132, 1986.

Andrews, G.C., and Domonkos, A.N.: Skin hemangioma and retrolental fibroplasia. Arch. Dermatol. 68:320, 1953.

Anonymous: Lattes RS, editor. Tumors of the soft tissues (revised). 2nd (series) ed. Washington, DC: Armed Forces Institute of Pathology, 1982.

Antonelli-Orlidge, A., Saunders, K.B., Smith, S.R., and D'Amore, P.A.: An activated form of transforming growth factor beta is produced by cocultures of endothelial cells and pericytes. Proc. Natl. Acad. Sci. U.S.A. 86: 4544–4548, 1989a.

Antonelli-Orlidge, A., Smith, S.R., and D'Amore, P.A.: Influence of pericytes on capillary endothelial cell growth. Am. Rev. Respir. Dis. 140:1129–1131, 1989b.

Atkinson, J.B., Mahour, G.H., Isaacs, H., and Ortega, J.A.: Hemangiopericytoma in infants and children: A report of six patients. Am. J. Surg. 148: 372–374, 1984.

Autio-Harmainen, H., Karttunen, T., Apaja-Sarkkinen, M., Dammert, K., and Ristell, L.: Laminin and type IV collagen in different histological stages of Kaposi's sarcoma and other vascular lesions of blood vessel or lymphatic vessel origin. Am. J. Surg. Pathol. 12:469–476, 1988.

Bailey, P.V., Weber, T.R., Tracy, T.F., O'Conner, D.M., and Sotelo-Avila, C.: Congenital hemangiopericytoma: An unusual vascular neoplasm of infancy. Surgery 114:936–941, 1993.

Barsky, S.H., Rosen, S., Greer, D.E., and Noe, J.M.: The nature and evolution of port wine stains: A computer-assisted study. J. Invest. Dermatol. 74:154–157, 1980.

Barter, R.H., Letterman, G.S., and Schurter, M.: Hemangiomas in pregnancy. Am. J. Obstet. Gynecol. 87:625–634, 1963.

Bean, W.B.: *Anonymous Vascular Spiders and Related Lesions of the Skin.* Springfield, IL: Charles C. Thomas, 1958.

Berard, M., Sordello, S., Ortega, N., Carrier, J.L., Peyri, N., Wassef, M., et al: Vascular endothelial growth factor confers a growth advantage in vitro and in vivo to stromal cells cultured from neonatal hemangiomas. Am. J. Pathol. 150:1315–1326, 1997.

Bircher, A.J., Koo, J.Y., Frieden, I.J., and Berger, T.G.: Angiodysplastic syndrome with capillary and venous malformation associated with soft tissue hypotrophy. Dermatology 189:292–296, 1994.

Bjorklund, H., Dalsgaard, C.J., Jonsson, C.E., and Hermansson, A.: Sensory and autonomic innervation of non-hairy and hairy human skin: An immunohistochemical study. Cell Tissue Res. 243:51–57, 1986.

Bluefarb, S.M., and Adams, L.A.: Arteriovenous malformation with angiodermatitis: Stasis dermatitis simulating Kaposi's sarcoma. Arch. Dermatol. 96:176–181, 1967.

Boon, L.M., Mulliken, J.B., Vikkula, M., Watkins, H., Seidman, J., Olsen, B.R., et al.: Assignment of a locus for dominantly inherited venous malformations to chromosome 9p. Hum. Mol. Genet. 3:1583–1587, 1994.

Boukobza, M., Enjolras, O., Guichard, J.P., Gelbert, F., Herbreteau, D., Reizine, D.: Cerebral developmental venous anomalies associated with head and neck venous malformations. Am. J. Neuroradiol. 17:987–994, 1996.

Bowers, R.E., Graham, E.A., and Thominson, K.M.: The natural history of the strawberry nevus. Arch. Dermatol. 82:667–670, 1960.

Bowman, C.A., Witte, M.J., Witte, C.L., Way, D.L., Nagle, R.B., Copeland, J.G., et al.: Cystic hygroma reconsidered: Hamartoma or neoplasm? Primary culture of an endothelial cell line from a massive cervicomediastinal hygroma with bony lymphangiomatosis. Lymphology 17:15–22, 1984.

Breier, G., Albrecht, U., Sterrer, S., and Risau, W.: Expression of vascular endothelial growth factor during embryonic angiogenesis and endothelial cell differentiation. Development 114:521–532, 1992.

Burns, A.J., Kaplan, L.C., and Mulliken, J.B.: Is there an association between hemangioma and syndromes with dysmorphic features? Pediatrics 88:1257–1267, 1991.

Byard, R.W., Burrows, P.E., Izakawa, T., et al.: Diffuse infantile hemangiomatosis: Clinicopathologic features and management problems in five fatal cases. Eur. J. Pediatr. 150:224–227, 1991.

Calonje, E., Mentzel, T., and Fletcher, C.D.: Pseudomalignant neural invasion in cellular ("infantile") capillary haemangiomas. Histopathology 26:159–164, 1995.

Calonje, E., and Wilson-Jones, E.: Vascular tumors: Tumors and tumor-like conditions of blood vessels and lymphatics. In Elder, D., Elenitsas, R., Jaworsky, C., and Johnson, B. (eds.): Lever's Histopathology of the Skin. 8th Ed. Philadelphia: Lipincott-Raven, 1997, p. 899.

Cancilla, P.A., Baker, R.N., Pollock, P.S., and Frommes, S.P.: The reaction of pericytes of the central nervous system to exogenous protein. Lab. Invest. 26:376–383, 1972.

Chen, K.T., Kassel, S.H., and Medrano, V.A.: Congenital hemangiopericytoma. J. Surg. Oncol. 31:127–129, 1986.

Chung, K.C., Weiss, S.W., and Kuzon, W.M.J.: Multifocal congenital hemangiopericytomas associated with Kasabach-Merritt syndrome. Br. J. Plast. Surg. 48:240–242, 1995.

Claman, H.N.: On scleroderma: Mast cells, endothelial cells, and fibroblasts. JAMA 262:1206–1209, 1989.

Clearkin, K.P., and Enzinger, F.M.: Intravascular papillary endothelial hyperplasia. Arch. Pathol. Lab. Med. 100:441–444, 1976.

Coffin, C.M.: Vascular tumors. In Coffin, C.M., Dehner, L.P., and O'Shea, P.A. (eds.): Pediatric Soft Tissue Tumors: A Clinical, Pathological, and Therapeutic Approach. Baltimore: Williams & Wilkins, 1997, pp. 40–79.

Coffin, C.M., and Dehner, L.P.: Fibroblastic-myofibroblastic tumors in children and adolescents: A clinicopathologic study of 108 examples in 103 patients. Pediatr. Pathol. 11:569–588, 1991.

Coffin, C.M., and Dehner, L.P.: Vascular tumors in children and adolescents: A clinicopathologic study of 228 tumors in 222 patients. Pathol. Annu. 28 (pt 1):97–120, 1993.

Cooper, A.G., and Bolande, R.P.: Multiple hemangiomas in an infant with cardiac hypertrophy: Mortem angiographic demonstration of the avenouse fistulae. Pediatrics 35:27–33, 1965.

Cotran, R.S.: Cellular components of inflammation: Endothelial cells. In Kelley, W.N., et al. (eds.): *Textbooks of Rheumatology.* 3rd Ed. Philadelphia: W.B. Saunders, 1989.

Davis, S., Aldrich, T.H., Jones, P.F., Acheson, A., Compton, D.L., Jain, V., et al.: Isolation of angiopoietin-1, a ligand for the TIE2 receptor, by secretion-trap expression cloning. Cell 87:1161–1169, 1996.

de Takats G: Vascular anomalies of the extremities. Surg. Gynecol. Obstet. 55:227, 1932.

Dethlefsen, S.M., Mulliken, J.B., and Glowacki, J: An ultrastructural study of mast cell interactions in hemangiomas. Ultrast. Pathol. 10:175–183, 1986.

Dumont, D.J., Anderson, L., Breitman, M.L., and Duncan, A.M: Assignment of the endothelial-specific protein receptor tyrosine kinase gene (TEK) to human chromosome 9p21. Genomics 23:512–513, 1994.

Earhart, R.N., Aeling, J.A., Nuss, D.D., and Mellette, J.R.: Pseudo-Kaposi sarcoma: A patient with arteriovenous malformation and skin lesions simulating Kaposi sarcoma. Arch. Dermatol. 110:907–910, 1974.

Emery, P.J., Bailey, C.M., and Evans, J.N.: Cystic hygroma of the head and neck. A review of 37 cases. J. Laryngol. Otol. 98:613–619, 1984.

Enjolras, O., Wassef, M., Mazoyer, E., Frieden, I.J., Rieu, P., Drouet, L., et al.: Infants with Kasabach-Merritt syndrome do not have "true" hemangiomas. J. Pediatr. 130:631–640, 1997.

Enzinger, F.M., and Smith, B.H.: Hemangiopericyotma: An analysis of 106 cases. Hum. Pathol. 7:61, 1976.

Esterly, N.B.: Cutaneous hemangiomas, vascular stains, and associated syndromes. Curr. Probl. Pediatr. 17:1–69, 1987.

Esterly, N.B., Margileth, A.M., Kahn, G., et al.: The management of disseminated eruptive hemangiomata in infants. Pediatr. Dermatol. 1:312–317, 1984.

Ezekowitz, R.A., Mulliken, J.B., and Folkman, J.: Interferon alfa-2a therapy for life-threatening hemangiomas of infancy. N. Engl. J. Med. 326:1456–1463, 1992.

Finley, J.L., Clark, R.A., Colvin, R.B., Blackman, R., Noe, J., and Rosen, S.: Immunofluorescent staining with antibodies to factor VIII, fibronectin, and collagenous basement membrane in normal human skin and port wine stains. Arch. Dermatol. 118:971–975, 1982.

Finley, J.L., Noe, J.M., Arndt, K.A., and Rosen, S.: Port-wine stains. Morphologic variations and developmental lesions. Arch. Dermatol. 120: 1453–1455, 1984.

Folkman, J.: Proceedings: Tumor angiogenesis factor. Cancer Res 34:2109–2113, 1974.

Folkman, J., and Cotran, R.S.: Relation of vascular proliferation to tumor growth. Int. Rev. Exp. Pathol. 16:207–248, 1976.

Folkman, J., and D'Amore, P.A.: Blood vessel formation: What is its molecular basis? Cell 87:1153–1155, 1996.

Folkman, J., and Klagsburn, M.: Angiogenic factors. Science 235:442–447, 1987.

Freisel, R., Komoriya, A., and Maciag, T.: Inhibition of endothelial cell proliferation by gamma-interferon. J. Cell. Biol. 104:689–696, 1987.

Fujimoto, T., and Singer, S.J.: Immunocytochemical studies of desmin and vimentin in pericapillary cells of chicken. J. Histochem. Cytochem. 35:1105–1115, 1987.

Gallione, C.J., Pasyk, K.A., Boon, L.M., Lennon, F., Johnson, D.W., Helmbold, E.A., et al.: A gene for familial venous malformations maps to chromosome 9p in a second large kindred. J. Med. Genet. 32:197–199, 1995.

Glick, S.A., Markstein, E.A., and Herreid, P.: Congenital glomangioma: Case report and review of the world literature. Pediatr. Dermatol. 12:242–244, 1995.

Glowacki, J., and Mulliken, J.B.: Mast cells in hemangiomas and vascular malformations. Pediatrics 70:48–51, 1982.

Golitz, L.E., Rudikoff, J., and O'Meara, O.P.: Diffuse neonatal hemangiomatosis. Pediatr. Dermatol. 3:145–152, 1986.

Gomes, M.M., and Bernatz, P.E.: Arteriovenous fistulas: A review and ten year experience at the Mayo Clinic. Mayo Clin. Proc. 45:81–102, 1970.

Gonzalez-Crussi, F., and Reyes-Mugica, M.: Cellular hemangiomas ("hemangioendotheliomas") in infants. Light microscopic, immunohistochemical, and ultrastructural observations. Am. J. Surg. Pathol. 15:769–778, 1991.

Gordon, J.R., Burd, P.R., and Galli, S.J.: Mast cells as a source of multifunctional cytokines. Immunol. Today 11:458–464, 1990.

Gozal, D., Saad, N., Bader, D., Berger, A., Jaffe, M.: Diffuse neonatal haemangiomatosis: Successful management with high dose corticosteroids. Eur. J. Pediatr. 149:321–324, 1990.

Grabb, W.C., Dingman, R.O., Oneal, R.M., and Dempsey, P.D.: Facial hamartomas in children: Neurofibroma, lymphangioma, and hemangioma. Plast. Reconstr. Surg. 66:509–527, 1980.

Hashimoto, H., Daimaru, Y., and Enjoji, M.: Intravascular papillary endothelial hyperplasia. A clinicopathologic study of 91 cases. Am. J. Dermatopathol. 5:539–546, 1983.

Hawkins, R.A., Claman, H.N., Clark, R.A., and Steigerwald, J.C.: Increased dermal mast cell populations in progressive systemic sclerosis: A link in chronic fibrosis. Ann. Intern. Med. 102:182–186, 1985.

Heagerty, A.H., Rubin, A., and Robinson, T.W.: Familial tufted angioma. Clin. Exp. Dermatol. 17:344–345, 1992.

Held, J.L., Haber, R.S., Silvers, D.N., and Grossman, M.E.: Benign neonatal hemangiomatosis: Review and description of a patient with unusually persistent lesions. Pediatr. Dermatol. 7:63–66, 1990.

Holden, K.R., and Alexander, F.: Diffuse neonatal hemangiomatosis. Pediatrics. 46:411–421, 1970.

Hood, A.F., Kwan, T.H., Mihm, M.C.J., and Horn, T.D.: *Primer of Dermatopathology.* 2nd Ed. Boston: Little, Brown and Company, 1993.

Hultberg, B.M., and Svanholm, H.: Immunohistochemical differentiation between lymphangiographically verified lymphatic vessels and blood vessels. Virchows Arch. [A] Pathol. Anat. Histopathol. 414:209–215, 1989.

Hunt, L.W., Colby, T.V., Weiler, D.A., Sur, S., and Butterfield, J.H.: Immunofluorescent staining for mast cells in idopathic pulmonary fibrosis: Quantification and evidence for extracellular release of mast cell tryptase. Mayo Clin. Proc. 67:941–948, 1992.

Iemura, A., Tsai, M., Ando, A., Wershil, B.K., and Galli, S.J.: The *c-kit* ligand, stem cell factor, promotes mast cell survival by suppressing apoptosis. Am. J. Pathol. 144:321–328, 1994.

Ito, N., Kawata, S., Tsushima, H., Tamura, S., Kiso, S., Takami, S.: Increased circulating transforming growth factor beta 1 in a patient with giant hepatic hemangioma: Possible contribution to an impaired immune function. Hepatology 25:93–96, 1997.

Jackson, I.T., Carreno, R., Potparic, Z., and Hussain, K.: Hemangiomas, vascular malformations, and lymphovenous malformations: Classification and methods of treatment. Plast. Reconstr. Surg. 1993; 91:1216–1230.

Johnston, M.C.A.: Radioautographic study of the migration and fate of cranial neural crest. Anat. Rec. 156:143, 1966.

Jones, E.W., and Orkin, M.: Tufted angioma (angioblastoma): A benign progressive angioma, not to be confused with Kaposi's sarcoma or low-grade angiosarcoma. J. Am. Acad. Dermatol. 20:214–225, 1989.

Joyce, N.C., Haire, M.F., and Palade, G.E.: Contractile proteins in pericytes: Immunocytochemical evidence for the presence of two isomyosins in graded concentrations. J. Cell Biol. 100:1387–1395, 1985.

Kaplan, E.N.: Vascular malformation of the extremities. In Williams, H.B. (ed.): Symposium on Vascular Malformations and Melanotic Lesions. St. Louis: C.V. Mosby, 1983, p. 144.

Kasabach, H.H., and Merritt, K.K.: Capillary hemangioma with extensive purpura: Report of a case. Am. J. Dis. Child. 59:1063–1070, 1940.

Kauffman, S.L., and Stout, A.P.: Hemangiopericytoma in children. Cancer 13:695–710, 1960.

Kessler, D.A., Langer, R.S., Pless, N.A., and Folkman, J.: Mast cells and tumor angiogenesis. Int. J. Cancer 18:703–709, 1976.

Kimura, S.: Ultrastructure of so-called angioblastoma of the skin before and after soft x-ray therapy. J. Dermatol. 8:235–243, 1981.

Koch, A.E., Halloran, M.M., Haskell, C.J., Shah, M.R., and Polverini, P.J.: Angiogenesis mediated by soluble forms of E-selectin and vascular cell adhesion molecule-1. Nature 376:517–519, 1995.

Kohout, E., and Stout, A.P.: The glomus tumor in children. Cancer 14:555–566, 1961.

Kraling, B.M., Razon, M.J., Boon, L.M., Zurakowski, D., Seachord, C., Darveau, R.P., et al.: E-selection is present in proliferating endothelial cells in human hemangiomas. Am. J. Pathol. 148:1181–1191, 1996.

Kumakiri, M., Muramoto, F., Tsukinaga, I., Yoshida, T., Ohura, T., and Miura, Y.: Crystalline lamellae in the endothelial cells of a type of hemangioma characterized by the proliferation of immature endoethial cells and pericytes—Angioblastoma (Nakagawa). J. Am. Acad. Dermatol. 8:68–75. 1983.

Lai, F.M., Allen, P.W., Yuen, P.M., and Leung, P.C.: Locally metastasizing vascular tumor: Spindle cell, epithelioid or unclassified hemangioendothelioma. Am. J. Clin. Pathol. 96:660–663, 1991.

Lam, W.Y., Mac-Moune Lai, F., Look, C.N., Choi, P.C., and Allen, P.W.: Tufted angioma with complete regression. J. Cutan. Pathol. 21:461–466, 1994.

Landthaler, M., Braun-Falco, O., Eckert, F., Stolz, W., Dorn, M., and Wolff, H.H.: Congenital multiple plaquelike glomus tumors. Arch. Dermatol. 126:1203–1207, 1990.

Lawton, R.C., Tidrick, R.T., and Brintnall, E.S.: A clinicopthological study of multiple congenital arteriovenous fistulae of the lower extremities. Angiology 8:161, 1957.

Lindenauer, S.M.: Klippel-Trenaunay syndrome. Ann. Surg. 162:303, 1965.

Listrom, M.B., and Fenoglio-Preiser, C.M.: Does laminin immunoreactivity really distinguish between lymphatics and blood vessels. Surg. Pathol. 1:71–74, 1988.

MacMillan, A., Champion, R.H.: Progressive capillary haemangioma. Br. J. Dermatol. 85:492–493, 1971.

Malan, E.: Vascular Malformations (Angiodysplasias). Milan: Carl Erba Foundation, 1974.

Masson, P.: Hemangioendotheliome vegetante intravasculaire. Bull. Soc. Anat. (Paris) 93:517–532, 1923.

Marsch, W.C.: The ultrastructure of eruptive hemangioma ("pyogenic granuloma"). J. Cutan. Pathol. 8:144–145, 1981.

Martinez-Perez, D., Fein, N.A., Boon, L. M., and Mulliken, J.B.: Not all hemangiomas look like strawberries: Uncommon presentations of the most common tumor of infancy. Pediatr. Dermatol. 12:1–6, 1995.

Martin-Padura, I., de Castellarnau, C., Uccini, S., Pilozzi, E., Natali, P.G., Nicotra, M.R.: Expression of VE (vascular endothelial)–cadherin and other endothelial-specific markers in haemangiomas. J. Pathol. 175:51–57, 1995.

Meininger, C.J., Brightman, S.E., Kelly, K.A., and Zetter, B.R.: Increased stem cell factor release by hemangioma-derived endothelial cells. Lab. Invest. 72:166–173, 1995.

Meininger, C.J., and Zetter, B.R.: Mast cells and angiogenesis. Semin. Cancer Biol. 3:73–79, 1992.

Mekori, Y.A., Oh, C.K., and Metcalfe, D.D.: IL-3–dependent murine mast cells undergo apoptosis on removal of IL-3. Prevention of apoptosis by *c-kit* ligand. J. Immunol. 151:3775–3784, 1993.

Mentzel, T., Calonje, E., Nascimento, A.G., and Fletcher, C.D.: Infantile hemangiopericytoma versus infantile myofibromatosis. Study of a series suggesting a spectrum of infantile myofibroblastic lesions. Am. J. Surg. Pathol. 18:922–930, 1994.

Miettinen, M., Lehto, V.P., and Virtanen, I.: Glomus tumor cells: Evaluation of smooth muscle and endothelial cell properties. Virchows Arch. [B] Cell. Pathol. 43:139–149, 1983.

Mignatti, P., Tsuboi, R., Robbins, E., and Rifkin, D.B.: In vitro angiogenesis on the human amniotic membrane: Requirement for basic fibroblast growth factor-induced proteinases. J. Cell. Biol. 108:671–682, 1989.

Mihm, M.C., Jr., Soter, N.A., Dvorak, H.F., and Austen, K.F.: The structure of normal skin and the morphology of atopic eczema. J. Invest. Dermatol. 67:305–312, 1976.

Miki, H., and Matsumoto, R.: Angioblastoma: Report of three cases. Nippon Hifuka Gakkai Zasshi 72:733–740, 1962.

Montesano, R., Vassalli, J.D., Baird, A., Guillemin, R., and Orci, L.: Basic fibroblast growth factor induces angiogenesis in vitro. Proc. Natl. Acad. Sci. U.S.A. 83:7297–7301, 1986.

Morgan, A., and Evbuomwan, I.: Congenital haemangiopericytoma of the face with early distant metastasis. J.R. Coll. Surg. Edinb. 28:123–125, 1983.

Morris, M.: Lymphangioma circumscriptum. In Unna, P., et al. (eds.): International Atlas of Rare Skin Diseases. London: H.K. Lewis, 1889, pp. 1–4.

Mulliken J.B.: Classification of vascular birthmarks. In: Mulliken J.B., Young A.E. (eds). *Vascular Birthmarks: Hemangiomas and Malformations.* Philadelphia: WB Saunders, 24–37, 1988.

Mulliken, J.B., Glowacki J.: Hemangiomas and vascular malformations in infants and children: A classification based on endothelial characteristics. Plast. Reconstr. Surg. 69:412–422, 1982.

Muthukrishnan L., Warder E., McNeil P.L.: Basic fibroblast growth factor is efficiently released from a cytosolic storage site through plasma membrane disruptions of endothelial cells. J. Cellul. Physiol. 148:1–16. 1991.

Nakagawa K. Case report of angioblastoma of the skin. Nippon Hifuka Gakkai Zasshi 59:92–94, 1949.

Nguyen M., Strubel N., Bischoff J.: A role for sialyl Lewis-X/A glycoconjugates in capillary morphogenesis. Nature 365:267–269, 1993.

Nichols G.E., Gaffey M.J., Mills S.E., Weiss L.M.: Lobular capillary hemangioma: An immunohistochemical study including steroid hormone receptor status. Am. J. Clin. Pathol. 97:770–775, 1992.

Niedt G.W., Greco M.A., Wieczorek R., Blanc W.A., Knowles D.M.: Hemangioma with Kaposi's sarcoma like features: report of two cases. Pediatr. Pathol. 9:567–575, 1989.

Norrby K., Jakobsson A., Sorbo J.: Mast cell secretion and angiogenesis, a quantitative study in rats and mice. Virch. Arch. B. Cell. Pathol. 57:251–256, 1989.

Nozue, T., and Tsuzaki, M.: Further studies on distribution on neural crest cells in prenatal and postnatal development in mice. Okajimas Fol. Anat. Jpn. 51:131, 1974.

Oehlschlaegel, G., and Muller, E.: Zum Granuloma pyogenicum sive telangiectaticum als Sonderfall des capillaren Hamangioms. Arch. Klin. Exp. Dermatol. 218:126–157, 1964.

Ordonez, N.G., Mackay, B., El-Naggar, A.K., and Byers, R.M.: Congenital hemangiopericytoma. An ultrastructural, immunocytochemical, and flow cytometric study. Arch. Pathol. Lab. Med. 117:934–937, 1993.

Pack, G.T., and Miller, T.R.: Hemangiomas, classification, diagnosis, and treatment. Angiology 1:405, 1950.

Padilla, R.S., Orkin, M., and Rosai, J.: Acquired "tufted" angioma (progressive capillary hemangioma). A distinctive clinicopathologic entity related to lobular capillary hemangioma. Am. J. Dermatopathol. 9:292–300, 1987.

Pasyk, K.A., Cherry, G.W., Grabb, W.C., Sasaki, G. H.: Quantitative evaluation of mast cells in cellularly dynamic and adynamic vascular malformations. Plast. Reconstr. Surg. 73:69–75, 1984.

Pasyk, K.A., Grabb, W.C., and Cherry, G.W.: Cellular haemangioma. Light and electron microscopic studies of two cases. Virchows Arch. [Pathol. Anat.] 396:103–126, 1982.

Pasyk, K.A., Grabb, W.C., and Cherry, G.W.: Crystalloid inclusions in endothelial cells of cellular and capillary hemangiomas. A possible sign of cellular immaturity. Arch. Dermatol. 119:134–137, 1983.

Pearl, G.S., and Mathews, W.H.: Congenital retroperitoneal hemangioendothelioma with Kasabach-Merritt syndrome. South. Med. J. 72:239–240, 1979.

Perrone, T.: Vessel-nerve intermingling in benign infantile hemangioendothelioma. Hum. Pathol. 16:198–200, 1985.

Pins, M.R., Rosenthal, D.I., Springfield, D.S., and Rosenberg, A.E.: Florid extravascular papillary endothelial hyperplasia (Masson's pseudoangiosarcoma) presenting as a soft-tissue sarcoma. Arch. Pathol. Lab. Med. 117:259–263, 1993.

Powell, T.G., West, C.R., Pharoah, P.O., and Cooke, R.W.: Epidemiology of strawberry haemangioma in low birthweight infants. Brit. J. Dermatol. 116:635–641, 1987.

Prioleau, P.G., and Santa Cruz, D.J.: Lymphangioma circumscriptum following radical mastectomy and radiation therapy. Cancer 42:1989–1991, 1978.

Qu, Z., Liebler, J.M., Powers, M.R., Galey, T., Ahmadi, P., Huang, X., et al.: Mast cells are a major source of basic fibroblast growth factor in chronic inflammation and cutaneous hemangioma. Am. J. Pathol. 147:564–573, 1995.

Ramani, P., and Shah, A.: Lymphangiomatosis: Histologic and immunohistochemical analysis of four cases. Am J Surg Pathol 17:329–335, 1993.

Reese, A.B., and Blodi, F.C.: Retrolental fibroplasia. Am. J. Ophthalmol. 34:1, 1951.

Resnick S.D., Lacey, S., and Jones, G.: Hemorrhagic complications in a rapidly growing, congenital hemangiopericytoma. Pediatr. Dermatol. 10:267–270, 1993.

Risau, W., Sariola, H., Zerwes, H.G., Sasse, J., Ekblom, P., Kemler, R., et al.: Vasculogenesis and angiogenesis in embryonic-stem-cell–derived embryoid bodies. Development 102:471–478, 1988.

Robinson, D., Tieder, M., Halperin, N., Burshtein, D., and Nevo, Z.: Maffucci's syndrome—The result of neural abnormalities? Evidence of mitogenic neurotransmitters present in enchondromas and soft tissue hemangiomas. Cancer 74:949–957, 1994.

Rusin, L.J., and Harrell, E.R.: Arteriovenous fistula: Cutaneous manifestations. Arch. Dermatol. 112:1135–1138, 1976.

Schirren, C.G., Eckert, F., Kind, P.: Multiple disseminated glomus tumors (glomangioma). Immunohistochemical analysis and differentiation from viscerocutaneous hemangiomatosis. Hautarzt 44:457–461, 1993.

Schnyder, U.W., and Keller, R.: Zur Klinik und Histologie der Angiome. III. Mitteilung. Zur Histologie und Pathogenese der senilen Angiome. Arch. Dermatol. Syph. (Berlin) 198:333–342, 1954.

Seibert, J.J., Seibert, R.W., Weisenburger, D.S., and Allsbrook, W.: Multiple congenital hemangiopericytomas of the head and neck. Laryngoscope 88:1006–1012, 1978.

Sims, D.E.: The pericyte—A review. Tissue Cell 18:153–174, 1986.

Smith, S.S., and Basu, P.K.: Mast cells in corneal immune reaction. Can. J. Opthalmol. 5:175–183, 1970.

Smoller, B.R., and Apfelberg, D.B.: Infantile (juvenile) capillary hemangioma: A tumor of heterogeneous cellular elements. J. Cutan. Pathol. 20:330–336, 1993.

Smoller, B.R., and Rosen, S.: Port-wine stains: A disease of altered neural modulation of blood vessels? Arch. Dermatol. 122:177–179, 1986.

Spraker, M.K.: The vascular lesions of childhood. Dermatol. Clin. 4:79–87, 1986.

Stal, S., Hamilton, S., and Spira, M.: Hemangiomas, lymphangiomas, and vascular malformations of the head and neck. Otolaryngol. Clin. North Am. 19:769–796, 1986.

Stern, J.K.: Wolf, J.E., and Jarratt, M.: Benign neonatal hemangiomatosis. J. Am. Acad. Dermatol. 4:442–445, 1981.

Stout, A.P., and Lattes, R.S.: Tumors of the Soft Tissues. Washington, DC: Armed Forces Institute of Pathology, 1967.

Strutton, G., Weedon, D.: Acro-angiodermatitis: A simulant of Kaposi's sarcoma. Am. J. Dermatopathol. 9:85–89, 1987.

Suri, C., Jones, P.F., Patan, S., Bartunkova, S., Maisonpierre, P.C., Davis, S., et al.: Requisite role of anigopoietin-1, a ligand for the TIE2 receptor, during embryonic angiogenesis. Cell 87:1171–1180, 1996.

Suzuki, Y., Hashimoto, K., Crissman, J., Kanzaki, T., and Nishiyana, S.: The value of blood group-specific lectin and endothelial associated antibodies in the diagnosis of vascular proliferations. J. Cutan. Pathol. 13:408–419, 1986.

Swerlick, R.A., and Cooper, P.H.: Pyogenic granuloma (lobular capillary hemangioma) within port-wine stains. J. Am. Acad. Dermatol. 8:627–630, 1983.

Szilagyi, D.E., Smith, R.F., Elliot, J.P., and Hageman, J.H.: Congenital arteriovenous anomalies of the limbs. Arch. Surg. 111:423–429, 1976.

Takahashi, K., Mulliken, J.B., Kozakewich, H.P.W., Rogers, R.A., Folkman, J., and Ezekowitz, R.A.B.: Cellular markers that distinguish the phases of hemangioma during infancy and childhood. J. Clin. Invest. 93:2357–2364, 1994.

Troschke, A., Weyers, W., and Schill, W.B.: Multiple familial glomangioma. Hautarzt 44:731–734, 1993.

Tsai, M., Shih, L.S., Newlands, G.F.J., et al.: The rat *c-kit* ligand, stem cell factor, induces the development of connective tissue-type and mucosal mast cells *in vivo*: Analysis by anatomical distribution, histochemistry, and protease phenotype. J. Exp. Med. 174:125–131, 1991.

Tsang, W.Y.W., and Chan, J.K.C.: Kaposi-like infantile hemangioendothelioma: A distinctive vascular neoplasm of the retroperitoneum. Am. J. Surg. Pathol. 15:982–989, 1991.

Venkatachalam, M.A., and Greally, J.G.: Fine structure of glomus tumor: Similarity of glomus cells to smooth muscle. Cancer 23:1176–1184, 1969.

Vikkula, M., Boon, L.M., Carraway, K.L., Calvert, J.T., Diamonti, A.J., Goumnerov, B., et al.: Vascular dysmorphogenesis caused by an activating mutation in the receptor tyrosine kinase TIE2. Cell 87:1181–1190, 1996.

Waldo, E.D., Vuletin, J.C., and Kaye, G.I.: The ultrastructure of vascular tumors: Additonal observations and a review of the literature. Pathol. Ann. 12(pt 2):279–308, 1977.

Warner, J., and Wilson-Jones, E.: Pyogenic granuloma recurring with multiple satellites. A report of 11 cases. Br. J. Dermatol. 80:218–227, 1968.

Watson, W.L., and McCarthy, W.D.: Blood and lymph vessel tumors. Surg. Gynecol. Obstet. 71:569, 1940.

Weber, K., and Braun-Falco, O.: Ultrastructure of blood vessels in human granulation tissue. Arch. Dermatol. Forschung. 248:29–44, 1973.

Wegner, G.: Ueber Lymphangiome. Arch. Klin. Chir. 20:641, 1877.

Weinblatt, M.E., Kahn, E., and Kochen, J.A.: Hemangioendothelioma with intravascular coagulation and ischemic colitis. Cancer 54:2300–2304, 1984.

Weiss, S.W.: Pedal hemangioma (venous malformation) occurring in Turner's syndrome: An additional manifestation of the syndrome. Hum. Pathol. 19:1015–1018, 1988.

Whimster, I.W.: The pathology of lymphangioma circumscriptum. Br. J. Dermatol. 94:473–486, 1976.

Wood, W.S., and Dimmick, J.E.: Multiple infiltrating glomus tumors in children. Cancer 40:1680–1685, 1977.

Yasunaga, C., Sueshi, K., Ohgami, H., Suita, S., and Kawanami, T.: Hetero-geneous expression of endothelial cell markers in infantile hemangioen-dothelioma. Immunohistochemical study of two solitary cases and one multiple one. Am. J. Clin. Pathol. 91:673–681, 1989.

Young, A.E.: Pathogenesis of vascular malformations. In Mulliken, J.B., and Young, A.E. (eds.): *Vascular Birthmarks: Hemangiomas and Malformations.* Philadelphia: W.B. Saunders, 1988a, p. 109.

Young, A.E.: Venous and arterial malformations. In Mulliken, J.B., and Young, A.E. (eds.): *Vascular Birthmarks: Hemangiomas and Malformations.* Phila-delphia: W.B. Saunders, 1988b, pp. 210–212.

Zetter, B.R.: Angiogenesis. State of the art. Chest 93:159S–166S, 1988.

Zukerberg, L.R., Nickoloff, B.J., and Weiss, S.W.: Kaposiform hemangioen-dothelioma of infancy and childhood: An aggressive neoplasm associated with Kasabach-Merritt syndrome and lymphangiomatosis. Am. J. Surg. Pathol. 17:321–328, 1993.

Diagnostic Imaging of Congenital Vascular Lesions

CHARLES A. JAMES, M.D.

An organized approach to imaging of pediatric head and neck vascular lesions starts with an understanding of the clinical and pathological features of the various vascular lesions encountered. A practical framework for the classification of vascular lesions proposed by Mulliken and Glowacki (1982) separates hemangiomas from vascular malformations (arterial, capillary, venous, lymphatic). Hemangiomas are vascular tumors of infancy that have a proliferative phase of endothelial cell hyperplasia followed by subsequent involution throughout childhood. Vascular malformations result from errors of vascular morphogenesis, have normal endothelial mitotic activity, and grow commensurate with the patient throughout life (Mulliken and Glowacki, 1982). Plain radiographs may be useful when displaying extrinsic mass effect on the airway with larger vascular lesions or to

Hemangiomas and Vascular Malformations of the Head and Neck, Edited by Waner, M.D. and Suen, M.D.
ISBN 0471-17597-8 © 1999 Wiley-Liss, Inc.

demonstrate smaller intrinsic airway lesions such as subglottic hemangioma (Fig. 6.1). Osseous plain radiographic findings may be reactive
in nature (increased bone size, bowing, osteoporosis, cortical thickening) or may show intraosseous lytic destruction of bone (Bliznak and
Staple, 1974; Burrows et al., 1983; Chiras et al., 1990). Though bone
changes are well demonstrated on computed tomography (CT), detection of vascular lesions and defining complete extent of these lesions is
better determined on magnetic resonance imaging (MRI) (Fig. 6.2)
(Levine et al., 1986). MRI provides the greatest contrast between the lesion and the surrounding muscle groups, fascial planes, and surrounding blood vessels. MRI can image extent of vascular lesions into the
chest and deep to osseous structures that cannot be imaged completely
with ultrasound.

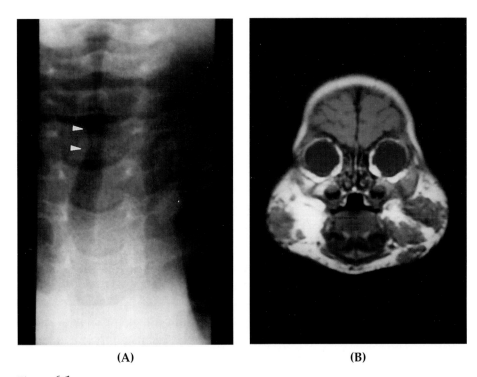

(A)   (B)

Figure 6.1
(A) Soft tissue cervical radiograph of an infant with stridor shows asymmetric subglottic narrowing of the trachea (arrowheads) due to subglottic hemangioma. (B)
Coronal T1-weighted MR scan of the head in the same infant shows multiple mass
lesions distinct from high signal intensity subcutaneous fat in this patient with
multifocal hemangiomas.

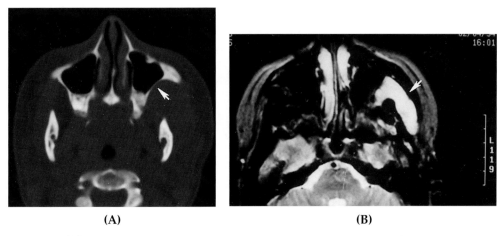

(A)   (B)

Figure 6.2
(A) Axial CT bone window image shows reactive bowing of the posterior wall of the left maxillary sinus (arrow) in this adolescent patient with an adjacent venous malformation. **(B)** Axial T2-weighted MRI scan fails to define the maxillary sinus bone changes. However, detection of the high signal intensity venous malformation (arrow) and distinction from adjacent structures is greater on the MRI scan than the CT images on soft tissue settings.

Continued improvement in ultrasound and MRI techniques aid in noninvasive distinction of high-flow lesions (proliferating heman-gioma, arteriovenous malformations) from slow-flow vascular lesions (venous malformations, lymphatic malformations). On ultrasound, a combination of gray scale features, color Doppler imaging and vascular waveform analysis is used to identify vascular lesions and to distinguish high-velocity/low-resistance arterial inflow of high-flow lesions from waveform characteristics of slow flow lesions (normal arterial flow volume, high arterial resistance) (Fig. 6.3) (Yakes, 1996). On spin echo MRI, low signal intensity flow voids are found in high-flow lesions. Increased sensitivity to high flow can be obtained by performing gradient recalled echo sequences with gradient moment nulling in which rapidly flowing protons appear as intense signal intensity (Meyer et al., 1991). Though the exact role of magnetic resonance angiography (MRA) is uncertain, the ability to provide a noninvasive angiogram may prove useful in planning invasive therapy or in providing noninvasive information following treatment. With improvements in these noninvasive imaging procedures, use of more invasive radiological procedures such as conventional angiography or direct punc-

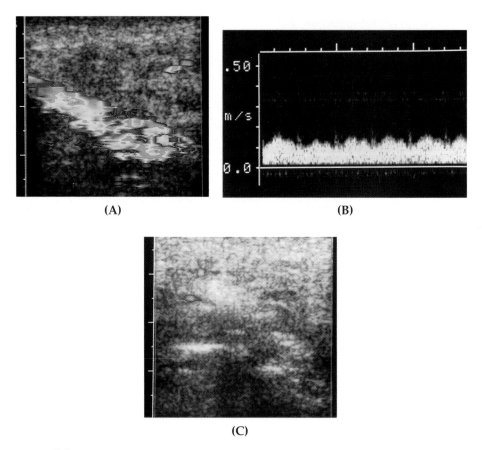

(A)   (B)

(C)

Figure 6.3
(A) Color flow ultrasound image demonstrates the vascular nature of the lesion in this infant with an extensive facial mass. (B) Doppler waveform analysis of the lesion shows a low resistance arterial waveform indicating high flow. (C) Post-treatment color flow ultrasound image shows resolution of a majority of the flow within the vascular lesion.

ture venography can usually be reserved for times when percutaneous treatment plans have been chosen.

6.1. HEMANGIOMAS

Hemangiomas are vascular lesions of infancy commonly encountered in the head and neck region as cutaneous or deep vascular masses with overlying (usually reddish) skin discoloration. Approximately 40% of lesions are present at birth, and presentation may be that of a single cutaneous lesion or multiple cutaneous lesions with multiple sites of organ system involvement (Mulliken and Glowacki, 1982; Ezekowitz et al., 1992). Rapid postnatal growth of hemangiomas is characteristic in the first 18 months of life during a proliferative phase followed by slow variable regression in lesion size over the first decade of life in the involutive phase. During the proliferate phase, endothelial cell hyperplasia noted histologically accounts for these soft tissue masses (Mulliken and Glowacki, 1982). In the involuting stage of hemangiomas, diminished cellularity within the lesion is apparent, and islands of fatty or fibrous tissue may be detected (Mulliken and Glowacki, 1982). A female predominance of hemangiomas is noted in contrast to vascular malformations, which show no female predominance and do not involute with growth of the patient (Mulliken and Glowacki, 1982; Lasjaunias and Berenstein, 1992). As uneventful involution of hemangiomas occurs in the majority of patients, conservative treatment is warranted. Less commonly, hemangiomas may impair functions such as vision or respiration based on critical lesion location; large or extensive lesions may be complicated by hemorrhage, platelet trapping, or high-output congestive heart failure (Ezekowitz et al., 1992; Lasjaunias and Berenstein, 1992). Within the head and neck, hemangiomas frequently involve the skin of the face, larynx, or orbit while sites such as the nasal cavity or retropharyngeal spaces are less commonly encountered (Ezekowitz et al., 1992). Radiographically, a soft tissue mass is generally seen; calcified phleboliths are usually absent, and underlying bone changes are absent or reactive (Burrows et al., 1983). CT may be helpful in rapid evaluation of widely extensive lesions (Fig. 6.4), though contrast between the lesion and surrounding structures is less than on MRI. With ultrasound, a well-defined focal mass with heterogeneous echotexture is usually seen (Yang et al., 1997). Color Doppler imaging with waveform analysis allows characterization of flow within the lesion and detection of arterial and venous waveforms relating to this vascular mass (Fig. 6.5). Though distinction of other head and neck masses (lymphadenopathy, congenital cysts, salivary gland tumors, and so forth) is usually possible, ultrasound is often limited in complete evaluation of lesion extent and in distinguishing tissue planes adjacent to the lesion (Yang et al., 1997).

Although angiography is less commonly performed because of advances in noninvasive imaging, it is helpful to understand angiographic findings in hemangiomas to better understand imaging findings

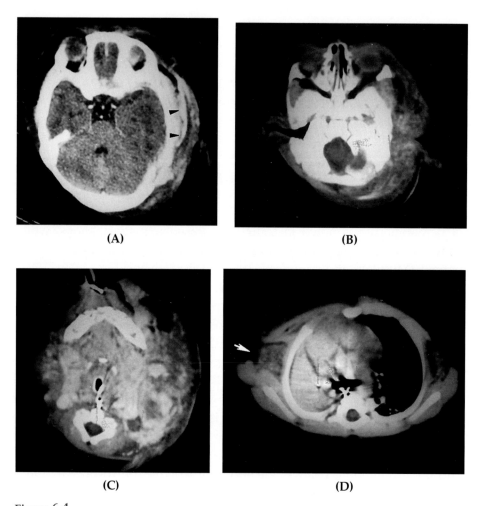

(A)   (B)

(C)   (D)

Figure 6.4
(A) Postintravenous contrast axial CT image of the head in an infant with an extensive cervicofacial vascular lesion with extent into the thorax. Extensive soft tissue thickening overlies the left temporal-occipital calvarium; an enhancing vessel relating to this lesion (arrowheads) indicates high flow. **(B)** More caudal CT image at the skull base shows an extensive infiltrative lesion of the subcutaneous tissues with obliteration of the left external auditory canal. **(C)** CT image in the neck shows prominent enhancing vessels relating to superficial and deep components of this proliferating hemangioma. Contrast discrimination between the vascular lesion and surrounding musculature is less than that provided on MRI. **(D)** Extent of the enhancing hemangioma into the right axillary soft tissues (arrow) is displayed. Right lung collapse is noted in this infant, requiring intubation due to cervical mass effect of the lesion.

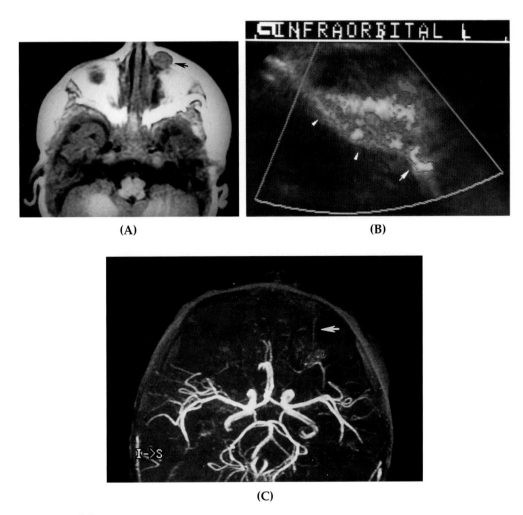

(A)   (B)

(C)

Figure **6.5**
(A) Axial T1-weighted MRI of a child with a left infraorbital mass shows a well-defined moderate signal soft tissue nodule (arrow) distinct from high signal subcutaneous fat. **(B)** Color flow ultrasound image shows flow throughout the infraorbital mass (arrowheads) and depicts an associated feeding vessel (arrow). **(C)** Time-of-flight MRA shows the feeding artery (arrow) supplying the hemangioma.

on ultrasound or MRI. At angiography, an organized pattern of arterial supply from adjacent arteries is evident in hemangiomas (Lasjaunias and Berenstein, 1992). An intense lobular parenchymal stain of well-circumscribed mass is seen (Burrows et al., 1983). The supplying arteries may be slightly enlarged and mildly tortuous; direct arteriovenous shunting is usually not demonstrated (Burrows et al., 1983). Other than dilation, the regional veins associated with the lesion are normal (Lasjaunias and Berenstein, 1992; Burrows et al., 1983). Angiography in the involuting phase of hemangiomas shows less intense tissue staining (Burrows et al., 1983).

MRI in the proliferative phase of hemangiomas will show findings of a soft tissue mass or masses with associated features of high flow (Fig. 6.6) (Meyer et al., 1991; Barnes et al., 1994). The solid tissue component is intermediate in signal intensity on T1-weighted sequences and markedly increased in signal intensity on T2-weighted imaging (Meyer et al., 1991; Baker et al., 1993). Vascular flow signal voids are seen relating to the mass on spin echo sequences with presaturation. The same vascular signal intensities appear as high-signal intensity flow enhancement on gradient echo sequences with gradient moment nulling (Meyer et al., 1991; Barnes et al., 1994). Prominent enhancement of the soft tissue component of the hemangioma is expected after the administration of intravenous gadolinium and helps differentiate the lesion from surrounding edematous changes associated with infiltrative lesions (Fig. 6.7) (Meyer et al., 1991; Baker et al., 1993). Non-enhancing areas within the soft tissue component of hemangiomas may be seen and correlate with sites of internal thrombosis within the lesion (Meyer et al., 1991). Delineation of deep cutaneous or intramuscular extent and detection of adjacent or remote satellite lesions may be seen on MRI (Baker et al., 1993). High-flow proliferating hemangiomas can be distinguished from high-flow arteriovenous malformations in that the latter lack the parenchymal component associated with hemangiomas (Barnes et al., 1994). Interval decrease of MRI findings of high-flow following treatment of hemangiomas may be seen (Meyer et al., 1991). Late involuting or involuted hemangiomas have features that overlap the low-flow vascular malformations such as venous malformations (Barnes et al., 1994). Correlation of imaging findings with prior clinical behavior of the lesion is imperative in such cases. Focal increased T1 signal intensity at sites of fatty replacement in partially involuted hemangiomas has been reported (Baker et al., 1993).

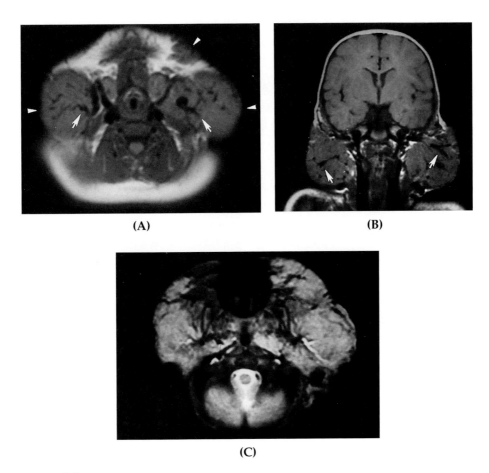

(A)   (B)

(C)

Figure 6.6

(A) Axial T1-weighted MRI of the neck in an infant with multiple cutaneous lesions with reddish skin discoloration. Multiple well-defined soft tissue masses (arrowheads) are seen and are well distinguished from adjacent high signal intensity fat. Note markedly enlarged vascular flow signal voids (arrows) within the larger lesions, indicating the high-flow nature of the hemangiomas. (B) Coronal T1-weighted MRI shows the bilateral moderate signal hemangioma masses. Branching vascular high-flow signal voids (arrows) are present within the lesions. (C) Axial T2-weighted MRI shows high signal intensity throughout the extensive hemangioma masses compared with lower T2-weighted signal intensity of subcutaneous fat.

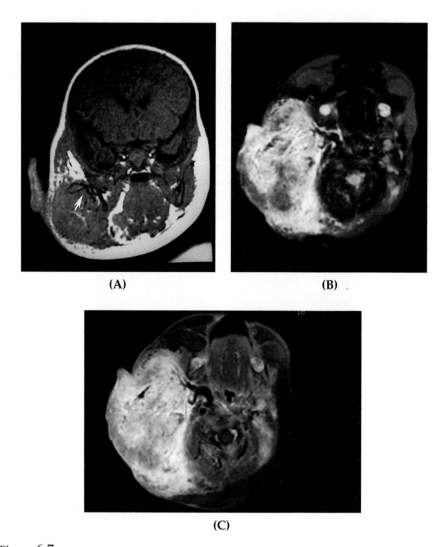

(A)   (B)

(C)

Figure 6.7
(A) Coronal T1-weighted MRI shows a large moderate signal mass distinguished from adjacent high signal intensity fat. Note prominent vascular flow signal voids (arrow) in this infant with clinical features of a proliferating hemangioma. **(B)** Axial T2-weighted MRI shows high signal intensity throughout the hemangioma mass lesion. The signal within the lesion is much greater than with subcutaneous fat, as expected on T2-weighted sequences. **(C)** Axial T1-weighted MRI following intravenous administration of gadolinium shows prominent enhancement throughout the proliferating hemangioma mass. Distinction of the enhancing lesion from subcutaneous fat is improved with fat-suppression techniques.

6.2. VENOUS MALFORMATIONS

Venous malformations are rare lesions resulting from abnormal developmental vein morphogenesis. Localized or diffuse ectatic, valveless venous spaces result with stagnant venous flow and variable communication with the surrounding normal venous system. Though present at birth, venous malformations generally are detected in late childhood or early adulthood as the lesion enlarges with growth of the patient (Yakes, 1996). The correct imaging approach of the lesion is made in conjunction with the clinical findings. Presentation of venous malformations ranges from an asymptomatic compressible soft tissue mass under normal or bluish skin to cosmetically deforming lesions that may interfere with functions such as mastication or respiration (Boukobza et al., 1996; Berthelsen et al., 1986). The lesions are compressible and increase in prominence with dependent positioning, crying, Valsalva maneuver, or obstruction of venous outflow (tourniquets) (Boukobza et al., 1996; Berthelsen et al., 1986; Boxt et al., 1983; Dubois et al., 1991). Lesion size or symptoms may increase in relation to exercise, menstruation, or pregnancy (Bliznak and Staple, 1974). The lesions may cause pain, interfere with muscle function, and be associated with hemorrhage at the skin or into deeper compartments (Boxt et al., 1983; Dubois et al., 1991; Bliznack and Staple, 1974; Yakes, 1994).

Extracranial venous malformations are rare but most commonly encountered in the head and neck or extremities with less frequent involvement of the trunk (Dubois et al., 1991; Rak et al., 1992). When involving the head and neck, facial lesions predominate over cervical lesions or orbital/frontal lesions (Boukobza et al., 1996). The temporomasseteric region and the cheek are common sites of involvement. Temporomasseteric lesions frequently extend into the parapharyngeal space (Gelbert et al., 1991). The lesions may involve multiple noncontiguous sites and tend to cross fascial planes commonly involving the face bilaterally (Boukobza et al., 1996; Dubois et al., 1991; Baker et al., 1993). The lesions may be subcutaneous or intramuscular or involve both spaces (Yang et al., 1997).

Plain radiographs may show a soft tissue mass or alteration of soft tissue planes and may frequently detect calcified phleboliths (Berthelsen et al., 1986; Boxt et al., 1983; Bliznak and Staple, 1974; Burrows et al., 1983). CT of venous malformations may display calcified phleboliths and secondary osseous changes (Fig. 6.8). Distinction

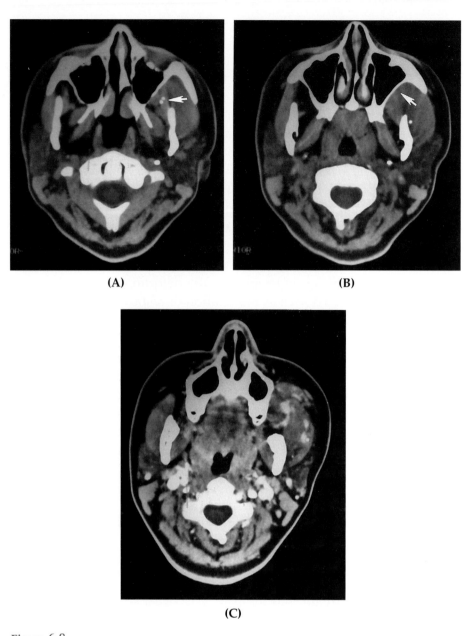

(A)   (B)

(C)

Figure 6.8
(A) Axial noncontrast CT examination shows two calcified phleboliths (arrow) within a venous malformation in this teenage boy. The lesion infiltrates fat planes lateral to the pterygoid muscles. Discrimination of the lesion from the left masseter muscle is uncertain on noncontrast CT image. (B) More caudal axial noncontrast CT image shows an additional calcified phlebolith within the venous malformation that mildly displaces the posterior wall of the left maxillary sinus (arrow). (C) Postintravenous contrast axial CT image shows partial heterogeneous enhancement within the luminal component of this venous malformation. Distinction of the lesion from the left masseter muscle remains uncertain on the postcontrast CT scan. (Case courtesy of Dr. Randolph Roberts, Munster, IN.)

of the lesion from adjacent structures and delineation of full extent of the lesion on CT imaging may be incomplete (Yang et al., 1997). Characterization of the lesion as a vascular mass may be indeterminate on CT imaging (Yoshida et al., 1995).

Because a venous malformation is a postcapillary lesion with extremity slow flow, angiography of the lesion is generally unimpressive. Normal inflow arterial size, an intact capillary bed, and lack of arteriovenous shunting is expected (Yakes, 1994). In the venous phase, slow partial filling and contrast pooling within ectatic dilated venous channels is often seen (Yakes, 1997; Burrows et al., 1983). The venous opacification may be lobular in configuration or may have a "striated" appearance, indicating intramuscular involvement (Burrows et al., 1983). Given dilution of contrast within the ectatic venous spaces, the lesion may have no opacification at angiography or may require superselective catheterization for detection of the venous abnormality (Berthelsen et al., 1986; Bliznak and Staple, 1974).

More complete opacification of the lesion is obtained by closed system venography or direct puncture venography. Closed system venography is used in the evaluation of the extent of extremity lesions during treatment planning. A tourniquet is used proximal to an extremity venous malformation while radiographic contrast is infused via routine venipuncture distal to the lesion (Fig. 6.9) (Geiser and Eversmann, 1978). By occluding venous outflow, much greater extent of opacification of the venous malformation occurs via sites of communication with the normal venous system. Filling of multiple clusters of tortuous, saccular components of venous malformation may be seen by this method in patients with subtle angiographic abnormalities (Bliznak and Staple, 1974; Geiser and Eversmann, 1978).

As the use of tourniquets is less feasible in the head and neck, direct percutaneous puncture of venous malformations evolved as a technique to opacify these lesions to better advantage (Berthelsen et al., 1986; Boxt et al., 1983). The lesion is punctured percutaneously through unaffected skin (Berthelsen et al., 1986). Upon venous return, the anomaly is filled with iodinated radiographic contrast under fluoroscopic control or digital subtraction techniques. Tortuous amorphous contrast collections pooling within these venous spaces are evident, and slow contrast washout should be expected (Fig. 6.9) (Boxt et al., 1983). The anomaly may consist of large venous cavities or a multiloculated cluster of interconnected venous spaces (Berthelsen et al., 1986; Dubois et

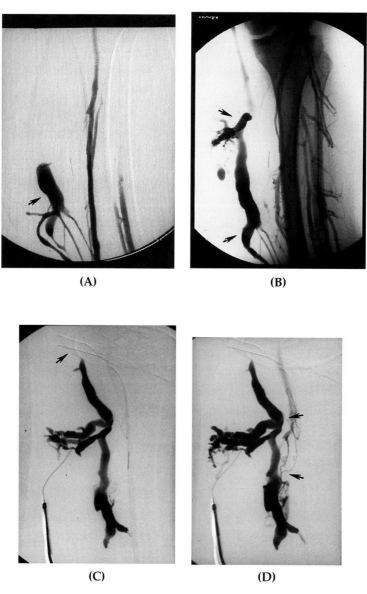

(A) **(B)**

(C) **(D)**

Figure 6.9
(A) Adult patient with left calf pain and occasional cutaneous bleeding relating to a venous malformation that has enlarged commensurate with growth of the patient. Early digital subtraction image from a closed system venogram shows partial filling of the normal deep venous system of the calf as well as early filling of the venous malformation (arrow). **(B)** Later unsubtracted image from the closed system venogram shows more complete contrast filling of the venous malformation (arrows) as well as increased filling of the normal deep venous system. **(C)** Direct puncture venogram image shows filling of the venous malformation prior to filling of the deep venous system. A tourniquet about the knee (arrow) is placed to control venous outflow from the malformation. **(D)** Subsequent image from direct puncture venogram shows more complete filling of the venous malformation and sites of venous communication with the deep venous system (arrows).

al., 1991). The volume of contrast required to fill the venous malformation and the routes of venous drainage present are evaluated. On direct puncture venography, these lesions may be well circumscribed with no visible draining vein, may drain into a normal-caliber venous system, or may drain into ectatic draining veins (Dubois et al., 1991).

The use of ultrasound at diagnosis in patients with venous malformations is generally to distinguish this low-flow venous lesion from other avascular soft tissue masses or from high-flow arteriovenous malformations. The lesion's appearance on gray scale ultrasound is that of a lesion with variable echogenicity with margins that may be well defined but are often irregular (Yang et al., 1997; Yoshida et al., 1995). If venous luminal elements predominate, hypoechoic cystic variceal venous spaces are seen that are partially compressible (Fig. 6.10). When cellular venous wall elements predominate, the lesion will appear more echogenic and be less compressible (Yakes, 1996). Echogenic phleboliths with acoustic shadowing may be seen. Color Doppler imaging with Doppler waveform analysis shows findings of slow flow with findings of normal arterial flow volume and high arterial resistance in nearby arteries (Yakes, 1996). Given marked slowing of flow in these lesions, maneuvers to augment flow in the lesion may be needed (Fig. 6.10). In a technique termed *auto-augmentation,* the venous malformation is compressed with the ultrasound transducer. Upon release of compression, rapid refilling of the lesion with venous blood improves detection of flow on color Doppler imaging (Yakes, 1996). Compared with MRI, ultrasound is limited in evaluation of the complete deep extent of lesions (Yang et al., 1997). In addition, ultrasound cannot image deep to osseous structures overlying a lesion.

When percutaneous sclerotherapy of venous malformations is performed, ultrasound may aid percutaneous imaging-guided cannulation of lesions not easily palpable on examination. After injection of a sclerosing agent into the lesion, noncompressible echogenic thrombus will appear in the malformation (Yakes et al., 1990). Ultrasound allows quick evaluation of portions of the malformation not thrombosed, aiding additional treatment at multiple sites in the same setting (Yakes, 1994). The painless noninvasive nature of ultrasound makes it ideal for subsequent follow-up for lesions after treatment (Fig. 6.10).

MRI is the ideal initial diagnostic imaging examination in venous malformations as the complete extent of the lesion is detected in a noninvasive fashion. MRI is more capable than CT in demonstrating

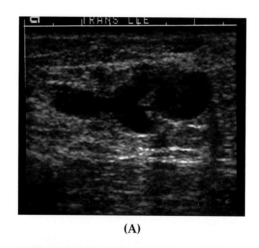

(A)

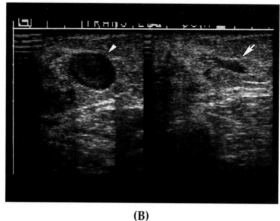

(B)

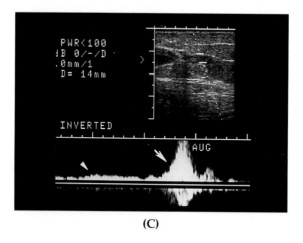

(C)

Figure 6.10
(A) Gray scale ultrasound image shows an irregular well-defined hypoechoic venous malformation. **(B)** Split screen transverse ultrasound imaging shows the venous malformation before compression (arrowhead) and near-complete collapse of the venous malformation on compression with the ultrasound transducer (arrow). This distinguishes the venous malformation from solid lesions and excludes luminal thrombosis within this portion of the lesion. **(C)** Doppler waveform analysis of the lesion shows a slow-flow venous waveform (arrowhead) that has increased venous flow (arrow) with augmentation maneuvers.

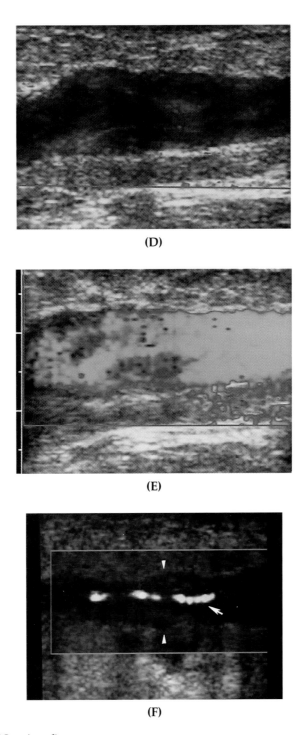

(D)

(E)

(F)

Figure **6.10** (*Continued*)
(D) Color Doppler ultrasound image without augmentation shows no appreciable flow within this slow flow venous malformation. **(E)** Color Doppler ultrasound image with augmentation techniques shows filling of the malformation with venous flow. **(F)** Ultrasound image following percutaneous sclerotherapy shows non-compressible echogenic thrombus filling the venous malformation (arrowheads). Only mild residual flow is detected (arrow) using augmentation techniques.

findings of high flow and in better defining the relationship of the malformation to adjacent muscles, organs, nerves, and tendons (Yakes,
1994). Compared with ultrasound, MRI displays a wider field of view,
can image lesion extent deep to osseous structures, and more often defines the full extent of lesions. Use of multiple orthogonal planes on
MRI aids in depicting relationship of the lesion to surrounding structures.

MRI features of venous malformations include signal less than fat
on T1-weighted imaging and signal increased to fat on T2-weighted
imaging (Fig. 6.11). The lesions have signal much higher than muscle
on T2-weighted images (Rak et al., 1992). If fast spin echo T2 imaging is
performed, fat suppression MR techniques should be used to increase
visualization of high signal intensity malformation from adjacent fat.
Venous lakes will show homogeneous high signal intensity on T2-
weighted imaging while the frequently present phlebolith will be low
signal foci on all pulse sequences (Baker et al., 1993). Associated bony
deformity, multifocal nature of a lesion, and associated subcutaneous
fatty prominence may be detected (Rak et al., 1992). Serpiginous signal
voids, characteristic of high-flow vascular lesions, are not seen (Baker

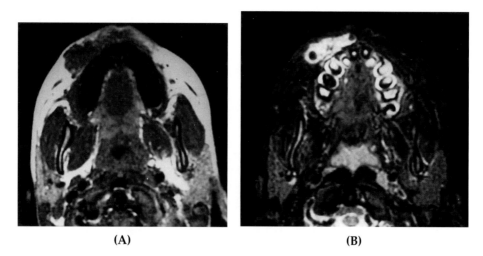

(A)   (B)

Figure 6.11
(A) Axial T1-weighted MRI shows a well-defined moderate signal intensity lesion
much less in signal intensity than surrounding fat. (B) Axial fast spin echo T2-
weighted MRI with fat suppression shows marked increased signal intensity of the
venous malformation compared with suppressed signal in adjacent fat. Oval low
signal intensities within the lesion represent calcified phleboliths.

et al., 1993). On gradient recalled echo imaging, the venous spaces are much less intense in signal than rapidly flowing protons of high-flow lesions (Fig. 6.12). After the administration of gadolinium, enhancement ranges from slight to prominent in a heterogeneous or homogeneous fashion (Fig. 6.13) (Gelbert et al., 1991; Baker et al., 1993; Meyer et al., 1991). The use of post-gadolinium MRI fat-suppression techniques will aid in the delineation of high-signal gadolinium enhancement within the lesion from high T1-weighted signal corresponding to postsclerotherapy thrombosis. Residual components of the lesion will remain high in signal intensity on T2 sequences post-treatment (Yakes, 1996).

6.3. ARTERIOVENOUS MALFORMATIONS

An arteriovenous malformation results when abnormal developmental morphogenesis of arterial, capillary, and venous components of the vascular system occurs in the same lesion. A loss in normal resistance at the capillary bed level results in arteriovenous shunting of this high-flow vascular malformation. In the head and neck, arteriovenous malformations occur less frequently than the low-flow vascular malformations (venous malformations, lymphatic malformations), though the clinical presentations may be more dramatic due to propensity of the lesion to cause significant hemorrhage (Erdmann et al., 1994; Des Prez et al., 1978; Lasjaunias and Berenstein, 1992). Head and neck vascular malformations commonly occur in the facial soft tissues near the oral cavity (cheek, nose, mandible), while sites such as the ear, palate, and scalp are less frequently involved (Erdmann et al., 1994; Des Prez et al., 1978; Gelbert et al., 1991).

Nonhemorrhagic presentations include facial deformity, pain, or functional disturbances (swallowing, breathing, vision) due to pressure or ischemic skin ulceration (Erdmann et al., 1994; Des Prez et al., 1978; Lasjaunias and Berenstein, 1992). The lesion may enlarge rather abruptly, and life-threatening hemorrhage may occur, particularly with teeth eruption or tooth extraction (Erdmann et al., 1994; Lasjaunias and Berenstein, 1992). On physical examination, a soft tissue fullness often underlies a region of increased warmth and discolored skin; dilated superficial veins may be apparent (Des Prez et al., 1978; Gelbert et al., 1991). A palpable thrill and audible bruit are expected in superficial lesions.

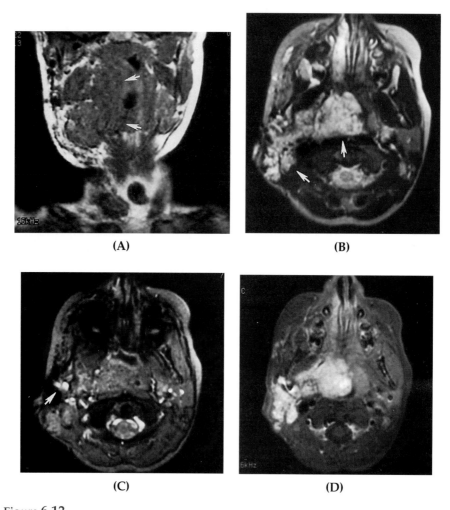

(A)   (B)

(C)   (D)

Figure 6.12
(A) Coronal T1-weighted MRI shows a large lobular superficial and deep lesion
(arrows) much lower in signal than overlying subcutaneous fat. (B) Axial T2-
weighted image shows high signal intensity throughout this extensive venous
malformation (arrows). Contrast discrimination between the lesion and the right
pterygoid muscles is greater than would be expected on CT imaging. (C) Axial
gradient-recalled echo MRI shows lack of high-flow intensity signal within this
slow-flow venous malformation lesion. Normal high signal intensity in adjacent
normal vessels is seen (arrow). (D) Axial post-gadolinium T1-weighted MRI
with fat suppression shows homogeneous lobular enhancement of this venous
malformation.

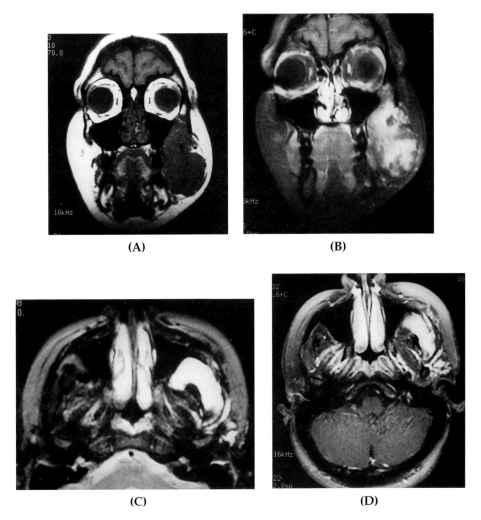

Figure 6.13
(A) Coronal T1-weighted MRI shows a large lobular moderate signal lesion much lower in signal than overlying fat. **(B)** Post-gadolinium coronal T1-weighted MRI with fat suppression shows enhancement of a majority of this venous malformation. **(C)** Axial T2-weighted image shows marked increased signal intensity of the deep component of the lesion relative to the lower signal of the pterygoid muscles and overlying fat. **(D)** Post-gadolinium axial T1-weighted image with fat suppression shows enhancement of this deep component of the venous malformation compared with nonenhancing adjacent musculature.

Plain radiographs may be normal, but a range of findings may be evident, including loss of soft tissue planes, asymmetrical bony overgrowth, cortical thickening, coarsened bony trabeculae, or discrete lytic destruction of bone (Des Prez et al., 1978; Cohen et al., 1986). Lytic lesions of bone may be unilocular or multilocular, and reossification of lytic lesions following treatment of arteriovenous malformations may be seen (Chiras et al., 1990).

Though CT accurately defines bone abnormalities of arteriovenous malformations (Fig. 6.14), discerning enhancement of vascular structures compared with adjacent soft tissue structures may be difficult (Yoshida et al., 1995). Additional limits of CT include requirement of intravenous contrast administration, limited transverse plane of data acquisition, and beam hardening artifact adjacent to bony structures (Cohen et al., 1986). With advances in color flow imaging, ultrasound can noninvasively characterize the lesion as vascular in nature, discerning the lesion from congenital cystic lesions or solid tumors of the head and neck. Gray scale imaging findings usually show a heterogeneous lesion with hypoechoic internal structures. Afferent and efferent vessels may be delineated (Yoshida et al., 1995; Yang et al., 1997). At color flow imaging, a mosaic pattern of vascular flow may be apparent within the tortuous vascular nidus (Fig. 6.15) (Yoshida et al., 1995). Waveform analysis quickly discerns arterial and venous components of the lesion and may show an arterialized waveform of a draining vein due to high-flow arteriovenous shunting (Yoshida et al., 1995; Yang et al., 1997). Ultrasound is limited in displaying the entire extent of vascular lesions, particularly in the delineation of deep soft tissue extent and the inability of ultrasound to image deep to bone.

Figure 6.14
(A) Axial CT image on bone window settings shows bone overgrowth of the right hemimandible relating to an adjacent arteriovenous malformation. **(B)** Postcontrast coronal CT image shows enhancing vessels (arrow) within an enlarged right masseter muscle. **(C)** More posterior postcontrast coronal CT image shows marked enlargement of a draining vein (arrow) relating to the arteriovenous malformation. (Case courtesy of Dr. C.W. McCluggage, Houston, TX.) **(D)** Post-gadolinium coronal T1-weighted MRI shows lack of a discrete enhancing mass. (Case courtesy of Dr. L.P. Gerson, Houston, TX.) **(E)** Lateral digital subtraction angiogram of the external carotid artery shows enlargement of feeding arteries (arrows) relating to the malformation. **(F)** Later angiogram image shows diffuse blush relating to this extensive malformation and early venous drainage (arrow) indicating arteriovenous shunting of blood flow. (Angiogram courtesy of Dr. M.E. Mawab, Houston, TX.)

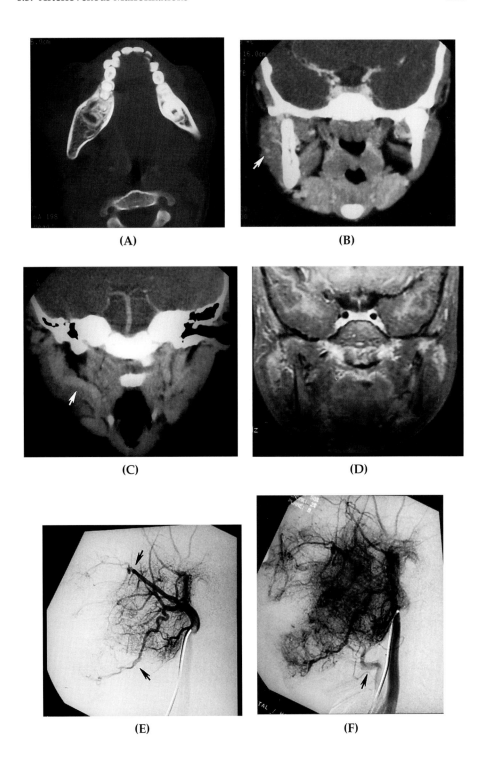

(A)

(B)

(C)

(D)

(E)

(F)

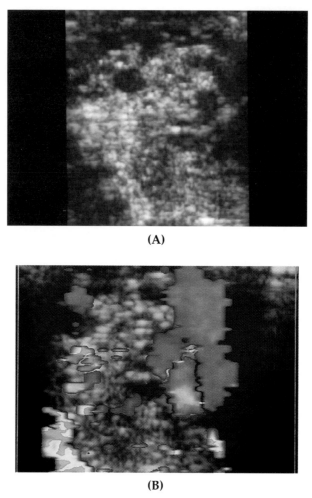

(A)

(B)

Figure 6.15
(A) Transverse gray scale ultrasound image shows a heterogeneous lesion with internal hypoechoic structures in this infant with a pulsatile lesion on palpation. (B) Color flow ultrasound image shows a mosaic pattern of tortuous flow within the arteriovenous malformation nidus.

MRI has been shown to reliably characterize arteriovenous malformations, to define extent of the lesion, and to distinguish the lesion from underlying muscle, bones, and tendons (Cohen et al., 1986; Meyer et al., 1991). The predominant MRI finding is that of prominent serpiginous flow voids on both T1 and T2 sequences relating to rapid flow in feeding arteries, malformation nidus, or draining veins (Fig. 6.16). Solid soft tissue abnormality of hemangioma or spin echo T2 high signal in-

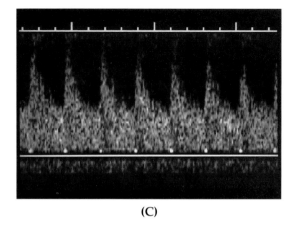

(C)

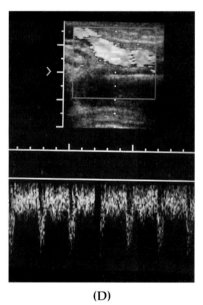

(D)

Figure **6.15** (*Continued*)
(**C**) Doppler waveform analysis within the nidus shows a high-velocity, low-resistance arterial waveform expected with arteriovenous shunting. (**D**) Doppler waveform evaluation of a markedly dilated draining vein shows arterialization of the venous waveform due to high-flow arteriovenous shunting

tensity of hemangiomas, venous malformations, or lymphatic malformations is absent on MR imaging of arteriovenous malformations (Fig. 6.17) (Gelbert et al., 1991; Meyer et al., 1991; Rak et al., 1992). Approximation of the anatomical area of involvement on MRI is obtained. Compared with CT imaging, MRI is particularly more sensitive

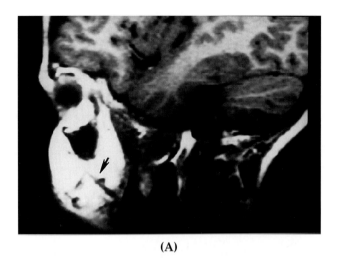

(A)

(B)

Figure 6.16
(A) Sagittal T1-weighted MRI shows serpiginous flow voids (arrow) in this child with a pulsatile lesion of the cheek. **(B)** Axial source images from time-of-flight MRA show high signal intensity within feeding vessels (arrows) relating to the arteriovenous malformation.

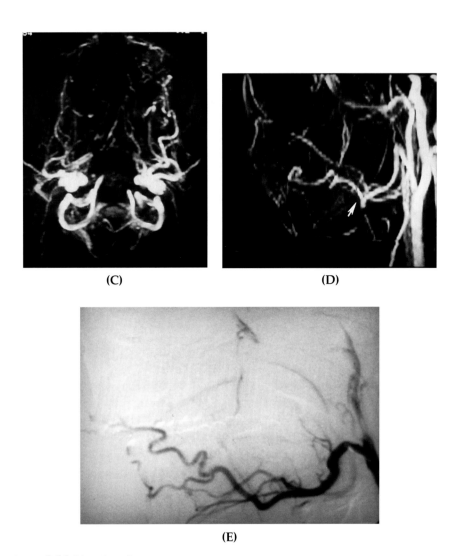

(C)   (D)

(E)

Figure 6.16 (*Continued*)
(C) Collapsed image from the MRA shows bilateral external carotid feeding vessels (left greater than right) extending to the left cheek arteriovenous malformation. (D) Lateral display image of MRA shows enlargement of external carotid artery branches, particularly the left facial artery (arrow). (E) Lateral digital subtraction angiogram of the left external carotid artery shows a markedly enlarged left facial artery feeding the arteriovenous malformation correlating with the prior MRA image.

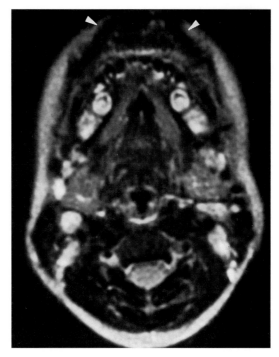

Figure **6.17**
Axial T2-weighted MRI shows a low signal intensity lesion (arrowheads) in this
patient with pathologically proven arteriovenous malformation. Other vascular lesions (hemangioma, venous malformation, lymphatic malformation) have significant components of high signal intensity on T2-weighted imaging.

at distinguishing muscle invasion from displacement (Gelbert et al.,
1991). Though major feeding arteries and draining veins are visualized,
the exact vessel involved is often uncertain (Cohen et al., 1986). In addition, delineation of feeding arteries and draining veins from the true
nidus of the malformation is difficult (Baker et al., 1993). Though low
signal intensity flow voids predominate, scattered punctate high signal
intensity may be seen due to flow phenomenon (e.g., flow-related enhancement, even-echo rephasing), thrombus, fat, or soft tissue injury
(Cohen et al., 1986). Though less conspicuous than on CT imaging, intraosseous involvement can be detected by decreased marrow signal,
especially on T1-weighted sequences (Baker et al., 1993).

Increased specificity for findings of high flow within lesions is obtained by complementing standard spin echo sequences with gradient

echo sequences in which MR imaging parameters are adjusted so that rapidly flowing protons appear as high signal intensity without administration of vascular contrast (Meyer et al., 1991; Rak et al., 1992). Lesions with prominent flow voids on spin echo images will show high signal intensity flow enhancement compared with stationary tissue (Meyer et al., 1991; Rak et al., 1992). More complex imaging sequences are used in MRA to actually perform a three-dimensional angiogram map without the invasive risks of conventional catheter angiography. In time-of-flight MRA, stationary tissues that remain in an imaging field are saturated by radiofrequency pulses. Moving tissue, such as flowing blood, enters the imaging field unsaturated and produces signal due to flow-related enhancement (Chien and Edelman, 1992). A volume of thin section source images is obtained, usually in the axial plane, when imaging the head and neck. Postprocessing of the dataset using maximum intensity projection (MIP) algorithm technique derives an image that can be rotated in any plane and that provides a three-dimensional angiogram map.

Use of MRA in vascular malformations has predominantly been in intracranial arteriovenous malformations where manipulation of the MRA and careful analysis of individual source images has been helpful in evaluating the supplying arteries, nidus, and draining veins or arteriovenous malformations (Angtuaco and Moran, 1997). MRA is more likely to identify specific vessels than routine spin echo MRI sequences, to correlate major feeders identified at angiography, and to display such vessels in multiple planes (Fig. 6.16). Analysis of individual data partitions is critical, and postprocessing techniques to remove overlapping vessels may improve details of information of the malformation. Though MRA may fail to identify all feeding vessels relating to an arteriovenous malformation, its noninvasive nature makes this modality well-suited for planning prior to invasive angiography and useful in follow-up imaging after embolic or surgical therapy.

An alternative to time-of-flight flow imaging is that of phase contrast MRA. In this technique, flow-related phase shifts are emphasized to provide signal separate from nearby stationary tissue (Chien and Edelman, 1992). This technique is more sensitive to slower flow than time-of-flight MRA but requires long scan times, which may be prohibitive when considering other multiple scan sequences used when MRI is employed to evaluate head and neck vascular lesions (T1, T2, postgadolinium T1, gradient echo sequences, and so forth).

Despite the progress in MRA, hemodynamic information of arteriovenous malformations is more exquisitely shown on selective conventional angiography. Flow patterns and directions of flow are directly shown in a dynamic fashion not capable on static MRA images (Angtuaco and Moran, 1997). Most head and neck arteriovenous mal-

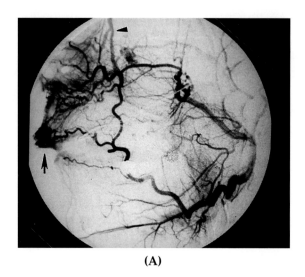

(A)

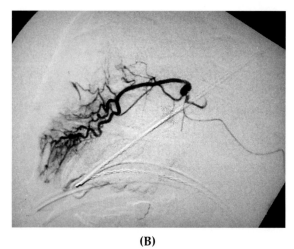

(B)

Figure 6.18
(A) Lateral digital subtraction angiogram of the external carotid artery shows enlarged feeding branches of the facial artery and the internal maxillary artery in this child with a pulsatile arteriovenous malformation of the upper lip and nasal region. Note nidus blush (arrow) and early draining veins (arrowhead). (B) Superselective microcatheter angiogram of the internal maxillary artery shows enlarged arterial branches supplying the arteriovenous malformation.

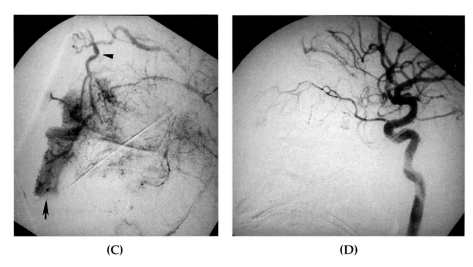

(C) (D)

Figure 6.18 (*Continued*)
(C) Later image from the internal maxillary angiogram run shows parenchymal blush at the arteriovenous malformation nidus (arrow) and early venous drainage via ophthalmic veins (arrowhead). (D) Lateral digital subtraction angiogram of the internal carotid artery shows small ophthalmic artery feeding branches extending to the region of the arteriovenous malformation.

formations are supplied by branches of the external carotid artery (Des Prez et al., 1978). Angiography should address overall topography of the lesion, delineate feeding arteries/draining veins, evaluate overall flow characteristics of the lesion, and assess for rapid arteriovenous shunting (Fig. 6.18) (Lasjaunias et al., 1992). Ectasia of both feeding arteries and draining veins is generally present. Evaluation of collateral contribution and arterial flow of adjacent normal tissue is imperative (Lasjaunias et al., 1992). Bilateral selective injection of all arterial pedicles in the territory of interest must be performed to provide an angiographic map of the entire lesion (Russell, 1986). Knowledge of the variable angiographic appearances of branches of the external carotid artery is imperative, and dangerous arterial flow patterns to the central nervous system, orbit, and cranial nerves must be elucidated to reduce the risks of stroke, blindness, or cranial nerve deficit should transcatheter embolic therapy be performed (Russell, 1986). Specifically, external carotid–internal carotid artery transosseous collateral routes or anomalous origin of the ophthalmic artery off the external carotid artery must be detected (Russell, 1986). Post-treatment angiography evaluates for change in flow pattern through the malformation (Chiras et al., 1990). If transcatheter embolization is performed, angiographic evaluation of the occluded artery must be followed by angiographic evaluation of nearby arterial branches to assess collateral flow contribution to any residual malformation (Russell, 1986).

6.4. LYMPHATIC MALFORMATIONS

Lymphatic malformations are slow-flow malformations of the lymphatic channels that may be localized or may be associated with generalized abnormality of the lymphatic system. When associated with a generalized lymphatic disorder, the presentation is generally that of hydrops and fetal demise in a fetus with an abnormal karyotype (usually Turner syndrome) and a posterior neck lymphatic malformation (Chervenak et al., 1983). In this scenario, imaging findings of hydrops (skin edema, ascites, pleural-pericardial effusions) are found on prenatal ultrasound in a fetus with posterior neck findings of lymphatic malformation (Chervenak et al., 1983). This section concerns the more localized lymphatic malformation of the head and neck that presents in infancy or childhood in patients in whom the remaining lymphatic system is normally formed. The malformation may involve only the lymphatic channels or may involve other vascular elements (e.g., lymphaticovenous malformations).

In traditional classification, the smallest lymphatic channel malformations were termed *capillary lymphangioma;* intermediate-sized lymphatic channel malformations were termed *cavernous lymphangioma,* while markedly dilated macrocystic lymphatic malformations were termed *cystic hygroma* (Zadvinskis et al., 1992).

These descriptive features may co-exist in a lymphatic malformation and are thought to represent the spectrum of developmental lymphatic malformation. In cystic hygroma, a primordial lymph sac probably fails to reestablish its communication with the central venous system. The predominance of this lesion at sites of loose connective tissue (e.g., axilla) may allow growth of the lesion to a large size (Zadvinskis et al., 1992). The smaller lymphatic channel malformations (cavernous/capillary lymphangiomas) may form from mesenchyme responsible for the more peripheral terminal lymphatic meshwork, and growth may be limited by tougher tissues (skin, muscle) where these smaller lymphatic malformations predominate (Zadvinskis et al., 1992).

A majority (80%–90%) of lymphatic malformations are detected in the first 2 years of life, and the neck is the most common site of involvement followed by the axilla (Zadvinskis et al., 1992; Emery et al., 1984). Within the neck, the posterior triangle is most frequently involved though the extent into or primary involvement of the submandibular region, floor of the mouth, cheek, parotid, and supraclavicular region may be encountered (Cohen and Thompson, 1986). Clinically, lymphatic malformations usually present as a painless nonpulsatile soft tissue mass, usually with overlying normal skin color (Emery et al., 1984; Cohen and Thompson, 1986; Nosan et al., 1995). Larger lesions tend to cross fascial planes, may be bilateral, and those in the anterior triangle may present with airway symptoms or dysphagia due to mass effect on the cervical esophagus or trachea (Emery et al., 1984). Extension of cervical lesions into the mediastinum occurs in up to 10% of cases, and respiratory distress due to intrathoracic displacement of the trachea may result (Sumner et al., 1981). Although most lymphatic malformations slowly enlarge over time, abrupt enlargement may occur due to complication of an existing lesion by internal hemorrhage or infection (Emery et al., 1984; Nosan et al., 1995). While adjacent organs are usually displaced by lymphatic malformations, infiltration of structures (e.g., parotid, tongue musculature, larynx) may occur.

Plain radiographs usually show a soft tissue mass and are most useful when large lesions cause displacement or mass effect on the airway (Sumner et al., 1981). Calcification within a lesion or bone destruction relating to a lymphatic malformation is extremely rare (Cohen and Thompson, 1986; Davidson and Hartman, 1990). Ultrasound is well suited for the evaluation of the superficial component of lymphatic malformations. In macrocystic lymphatic malformations the characteristic ultrasound appearance is that of a multiloculated cystic mass with intervening septa of variable thickness (Fig. 6.19) (Sheth et al., 1987). The cystic components are often hypoechoic but may be complicated with internal echogenic dependent hemorrhage or inflammatory debris (Davidson and Hartman, 1990; Ros et al., 1987). A discrete fluid-fluid level may be visualized, or the entire cystic component may be filled with echogenic debris or hemorrhage (Sheth et al., 1987; Ros et al., 1987). Lymphatic malformations may have ill-defined margins on ultrasound, and the lesion may be shown to infiltrate between tissue planes (Sheth et al., 1987).

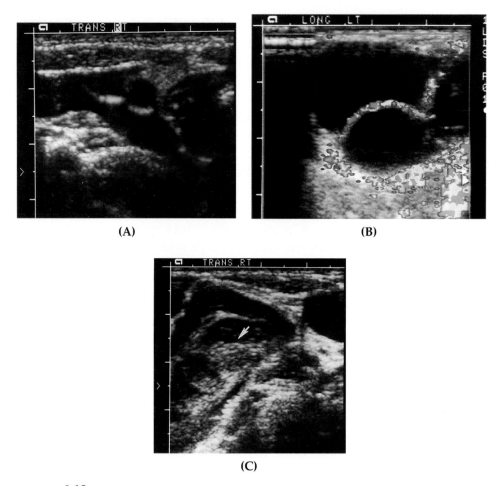

(A)   (B)

(C)

Figure 6.19
(A) Transverse ultrasound image of the neck shows a multiseptated predominantly hypoechoic cystic mass characteristic of a lymphatic malformation. (B)
Color flow ultrasound image shows hyperemia of the intervening septa. The cystic
portions of the lesion show no evidence of arterial or venous flow. (C) Transverse
ultrasound in a different portion of the lesion shows a dependent echogenic fluid
level (arrow) due to internal hemorrhage or inflammatory debris

Compared with CT, ultrasound is more sensitive in the detection of
the internal septations within cystic lesions, thereby favoring the diagnosis of lymphatic malformation over other cystic lesions of the head
and neck (Ros et al., 1987; Glasier et al., 1992). Ultrasound distinguishes
unilocular neck masses by showing characteristic alignment of
branchial cleft cysts (anterolateral to carotid sheath) or thyroglossal

duct cysts (midline near the hyoid) and demonstrates a thicker in-flamed wall of a soft tissue abscess. In addition to inflammatory debris or hemorrhage, echogenic components of lymphatic malformations in-clude clumps of solid, small lymphatic channels (microcystic lymphatic malformations), too small to resolve at ultrasound or, less commonly, echogenic calcified thrombus that may have deep acoustic shadowing (Sheth et al., 1987). Though the septations may be hyperemic on color Doppler imaging, a majority of the lesion is avascular. Therefore, ultra-sound distinguishes macrocystic lymphatic malformations from he-mangiomas, which are predominately echogenic (solid) and which have color Doppler imaging findings of flow throughout much of the lesion (Sheth et al., 1987; Yang et al., 1997). As microcystic lymphatic malformations have greater number of lymphatic septae, more color Doppler hyperemia can be expected.

A major limitation of ultrasound is failure to demonstrate the entire extent of many lymphatic malformations, particularly deep extent into the chest or deep to bony structures that obscure the acoustic window. In these instances, MRI or CT will be required for complete assessment of the lesion.

On postintravenous contrast CT examination, lymphatic malforma-tions are well-defined lesions of predominately homogeneous water at-tenuation (Fig. 6.20). The cystic components less commonly are hetero-geneous due to detection of internal hemorrhage or inflammatory debris (Fig. 6.21) (Davidson and Hartman, 1990). Rarely, a lesion will have internal fatty attenuation characterizing lymph from nonfatty cys-tic lesions (Davidson and Hartman, 1990; Ros et al., 1987). Though the lesion is predominantly avascular, the wall and septa of lymphatic malformations display variable thickness of enhancement on CT (Davidson and Hartman, 1990; Ros et al., 1987). While relationship to enhancing vessels is displayed, CT may struggle in the distinction of less cystic components of lymphatic malformations from other soft tis-sue masses or adjacent normal musculature (Siegel et al., 1989). Although osseous destruction is well displayed on CT examination, this finding will be encountered only rarely in head and neck lym-phatic malformations (Sinard and Welling, 1995).

With improved contrast between the lesion and surrounding soft tissues and ability to display the lesions in multiple imaging planes, MRI has distinct advantages over CT in providing the complete extent and pretreatment assessment of lymphatic malformations. On

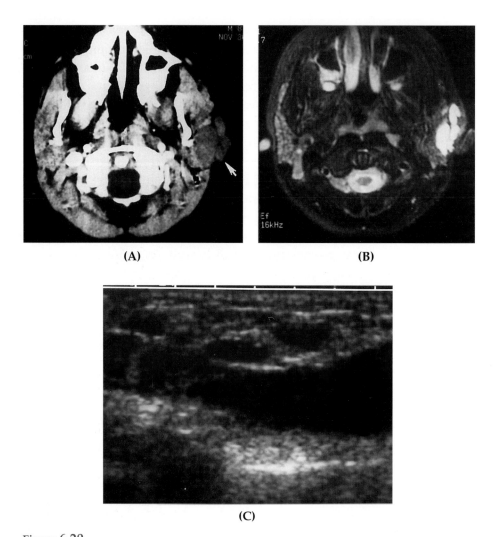

Figure 6.20

(A) Postcontrast CT scan shows a lobular water attenuation lesion (arrow) that infiltrates the left parotid gland. **(B)** Axial T2-weighted MRI shows high signal intensity within the lymphatic malformation and discriminates the lesion from the infiltrated moderate signal intensity parotid gland. **(C)** Transverse ultrasound image of the lesion shows a cystic lesion with internal variable thickness septations characteristic of lymphatic malformations.

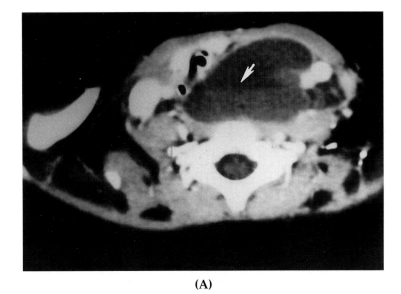

(A)

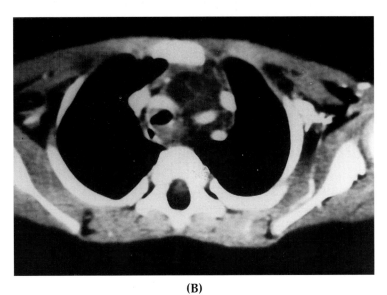

(B)

Figure **6.21**
(A) Postcontrast CT scan of the neck shows an infiltrative rim-enhancing cystic lesion with an internal hemorrhage level (arrow). Note mass effect on the trachea in
this adolescent patient who presented with abrupt onset of respiratory distress. **(B)**
Postcontrast CT image of the chest shows thoracic extent of the lesion. This appearance of a multiseptated water attenuation (cystic) mass that infiltrates fascial
planes is characteristic of lymphatic malformations.

T1-weighted MRI sequences, lymphatic malformations are usually similar to or slightly less than muscle in signal intensity. On T2-weighted sequences, the lesions have marked increased signal intensity, higher than fat and adjacent muscle (Fig. 6.22) (Siegel et al., 1989). In nearly one-fourth of lesions, T1 signal intensity of lymphatic malformations is near that of fat due to hemorrhage products or a higher component of fat in lesions with thick fibro-fatty septa between smaller caliber lymphatic channels (Fig. 6.23) (Siegel et al., 1989). Lymphatic malformations are usually heterogeneous in signal on both T1 and T2 weighted imaging and fluid-fluid levels are characteristic when seen. On T2 weighted imaging, low signal septations are usually seen coursing through a macrocystic lymphatic lesion; low signal on T2 may also relate to acute hemorrhage products (Fig. 6.24) (Siegel et al., 1989; Meyer et al., 1991). In some microcystic lymphatic malformations, components of the lesion may have relatively low signal on T2-weighted imaging due to greater lymphatic channel tissue components than the higher signal of adjacent cystic components (Fig. 6.25) (Gelbert et al., 1991). The margins are oval in shape and usually well defined but frequently have infiltrative borders (Siegel et al., 1989). Relationship of lymphatic malformations to adjacent vessels is shown without administration of intravenous contrast, which is an advantage over CT imaging. MRI findings depicting high flow are absent in lymphatic malformations. Gadolinium enhancement occurs in lymphatic malformations only in microcystic components of the lesion or if the lymphatic malformation occurs in conjunction with other malformations (e.g., lymphaticovenous malformations) (Meyer et al., 1991). As treatment of the lesion is often incomplete, imaging of recurrence of lymphatic malformations is frequently necessary.

Figure 6.22

(A) Coronal T1-weighted image shows a lobular moderate signal intensity lesion (arrowheads) that is isointense to muscle. (B) Axial T2-weighted image shows marked increased signal intensity of the lesion relative to adjacent fat and muscle. Note characteristic intervening septations coursing through the macrocystic lymphatic malformation. (C) Sagittal T2-weighted MRI shows extent of the lesion into the superior mediastinum. (A, B, reprinted from Seibert and James, 1997, with permission of the publisher.)

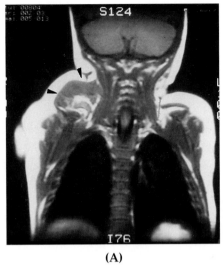

(A)

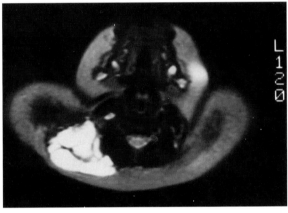

(B)

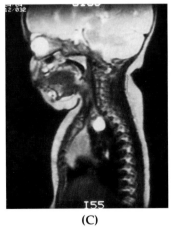

(C)

209

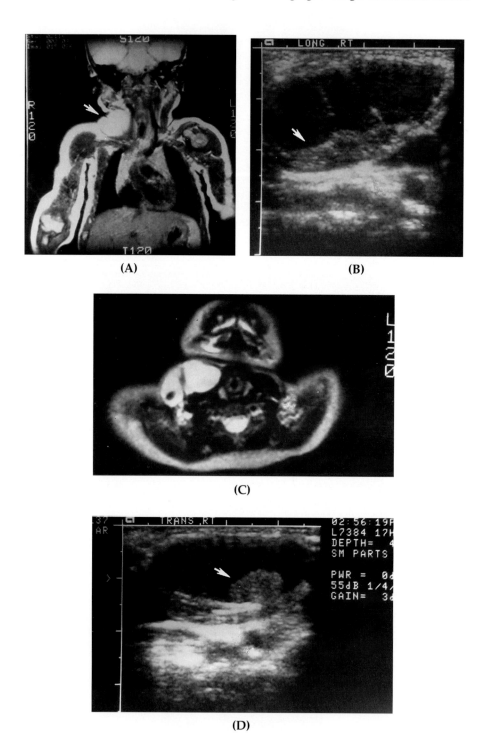

(A)

(B)

(C)

(D)

Figure 6.23
(A) Coronal T1-weighted MRI shows high signal intensity equal to fat in this lymphatic malformation (arrow), suggesting internal hemorrhage. (B) Longitudinal ultrasound image shows a multiseptated cystic lesion with dependent echogenic hemorrhage (arrow). (C) Axial T2-weighted MRI shows a septated lymphatic malformation predominantly of high signal intensity. An internal oval low signal focus represents thrombus but could be mistaken for a calcified phlebolith of a venous malformation. (D) Transverse ultrasound image shows a moderately echogenic dependent thrombus (arrow) in this cystic lymphatic malformation.

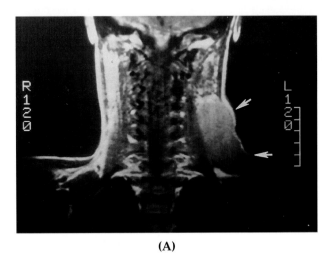

(A)

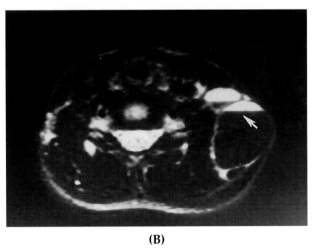

(B)

Figure 6.24
(A) Coronal T1-weighted MRI shows a lobular lymphatic malformation (arrows) with signal greater than adjacent muscle. (B) Axial T2-weighted MRI shows low signal intensity fluid-fluid levels (arrow) due to acute hemorrhage into this lymphatic malformation.

211

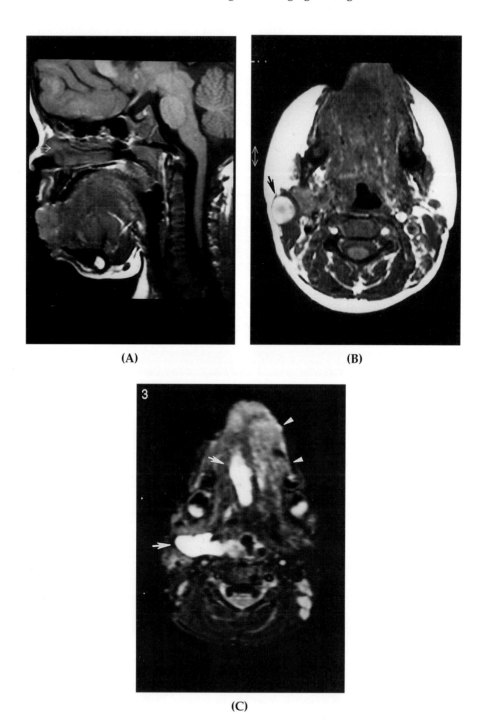

(A)   (B)

(C)

Figure 6.25
(A) Sagittal T1-weighted MRI shows a poorly defined infiltrative lesion that causes marked enlargement of the tongue and fills the oropharynx. **(B)** Axial T1-weighted MRI shows extent of this infiltrative lymphatic malformation into the right parotid gland. A component of the lesion in the parotid gland has high T1-signal intensity (arrow) probably due to internal hemorrhage. **(C)** Axial T2-weighted MRI shows characteristic high signal intensity within portions of the lymphatic malformation (arrows). Other components of the lesion have relatively lower signal intensity (arrowheads), probably due to greater lymphatic septal tissue components in this portion of the lesion. The tongue musculature and the parotid gland are common sites of lymphatic malformation infiltration. (Case courtesy of Dr. Nancy Rollins, Dallas, TX.)

REFERENCES

Angtuaco, E.E.C., and Moran, C.J.: MRA of vascular malformations. Clin. Neurosci. 4:130–136, 1997.

Baker, L.L., Dillon, W.P., Hieshima, G.B., Dowd, C.F., and Frieden, I.J.: Hemangiomas and vascular malformations of the head and neck: MR characterization. A.J.N.R. 14:307–314, 1993.

Barnes, P.D., Burrows, P.E., and Hoffer, F.A.: Hemangiomas and vascular malformations of the head and neck: MR characterization. A.J.N.R. 15:193–195, 1994.

Berthelsen, B., Fogdestam, I., and Svendsen, P.: Venous malformations in the face and neck: Radiologic diagnosis and treatment with absolute ethanol. Acta Radiol. Diagn. 27:149–155, 1986.

Bliznak, J., and Staple, T.W.: Radiology of angiodysplasias of the limb. Radiology 110:35–44, 1974.

Boukobza, M., Enjolras, O., Guichard, J.-P., et al.: Cerebral developmental venous anomalies associated with head and neck venous malformations. A.J.N.R. 17:987–994, 1996.

Boxt, L.M., Levin, D.C., and Fellows, K.E.: Direct puncture angiography in congenital venous malformations. A.J.R. 140:135–136, 1983.

Burrows, P.E., Mulliken, J.B., Fellows, K.E., and Strand, R.D.: Childhood hemangiomas and vascular malformations: Angiographic differentiation. A.J.R. 141:483–488, 1983.

Chervenak, F.A., Isaacson, G., Blakemore, K.J., et al.: Fetal cystic hygroma: Cause and natural history. N. Engl. J. Med. 309:822–825, 1983.

Chien, D., and Edelman, R.R.: Basic principles and clinical applications of magnetic resonance angiography. Semin. Roentgenol. 27:53–62, 1992.

Chiras, J., Hassine, D., Goudot, P., Meder, J.F., Guilbert, J.F., and Bories, J.: Treatment of arteriovenous malformations of the mandible by arterial and venous embolization. A.J.N.R. 11:1191–1194, 1990.

Cohen, J.M., Weinreb, J.C., and Redman, H.C.: Arteriovenous malformations of the extremities: MR imaging. Radiology 158:475–479, 1986.

Cohen, S.R., and Thompson, J.W.: Lymphangiomas of the larynx in infants and children: A survey of pediatric lymphangioma. Ann. Otol. Rhinol. Laryngol. 95:1–20, 1986.

Davidson, A.J., and Hartman, D.S.: Lymphangioma of the retroperitoneum: CT and sonographic characteristics. Radiology 175:507–510, 1990.

Des Prez, J.D., Kiehn, C.L., Vlastou, C., and Bonstelle, C.: Congenital arteriovenous malformation of the head and neck. Am. J. Surg. 136:424–429, 1978.

Dubois, J.M., Sebag, G.H., DeProst, Y., Teillac, D., Chretien, B., and Brunelle, F.O.: Soft-tissue venous malformations in children: Percutaneous sclerotherapy with ethibloc. Radiology 180:195–198, 1991.

Emery, P.J., Bailey, C.M., and Evans, J.N.G.: Cystic hygroma of the head and neck: A review of 37 cases. J. Laryngol. Otol. 98:613–619, 1984.

Erdmann, M.W.H., Davies, D.M., Jackson, J.E., and Allison, D.J.: Multidisciplinary approach to the management of head and neck arteriovenous malformations. Ann. R. Coll. Surg. Engl. 77:53–59, 1994.

Ezekowitz, R.A.B., Mulliken, J.B., and Folkman, J.: Interferon alfa-2a therapy for life-threatening hemangiomas of infancy. N. Engl. J. Med. 326:1456–1463, 1992.

Geiser, J.H., and Eversmann, W.W. Jr.: Closed system venography in the evaluation of upper extremity hemangiomas. J. Hand. Surg. 3:173–178, 1978.

Gelbert, F., Riche, M.C., Reizine, D., et al.: MR imaging of head and neck vascular malformations. J.M.R.I. 1:579–584, 1991.

Glasier, C.M., Brodsky, M.C., Leithiser, R.E. Jr., Williamson, S.L., and Seibert, J.J.: High resolution ultrasound with Doppler: A diagnostic adjunct in orbital and ocular lesions in children. Pediatr. Radiol. 22:174–178, 1992.

Lasjaunias, P., and Berenstein, A.: Craniofacial vascular lesions: General. In *Surgical Neuroangiography, II: Endovascular Treatment of Craniofacial Lesions.* Vol. 9. Heidelberg: Springer-Verlag, 1992, pp. 317–340.

Levine, E., Wetzel, L.H., and Neff, J.R.: MR imaging and CT of extrahepatic cavernous hemangiomas. A.J.R. 147:1299–1304, 1986.

Meyer, J.S., Hoffer, F.A., Barnes, P.D., and Mulliken, J.B.: Biological classification of soft-tissue vascular anomalies: MR correlation. A.J.R. 157:559–564, 1991.

Mulliken, J.B., and Glowacki, J.: Hemangiomas and vascular malformations in infants and children: A classification based on endothelial characteristics. Plast. Reconstr. Surg. 69:412–420, 1982.

Nosan, D.K., Martin, D.S., and Stith, J.A.: Lymphangioma presenting as a delayed post traumatic expanding neck mass. Am. J. Otolaryngol. 16: 186–189, 1995.

Rak, K.M., Yakes, W.F., Ray, R.L., et al.: MR imaging of symptomatic peripheral vascular malformations. A.J.R. 159:107–112, 1992.

Ros, P.R., Olmsted, W.W., Moser, R.P. Jr., Dachman, A.H., Hjermstad, B.H., and Sobin, L.H.: Mesenteric and omental cysts: Histologic classification with imaging correlation. Radiology 164:327–332, 1987.

Russell, E.J.: Functional angiography of the head and neck. A.J.N.R. 7:927–936, 1986.

Sheth, S., Nussbaum, A.R., Hutchins, G.M., and Sanders, R.C.: Cystic hygromas in children: Sonographic-pathologic correlation. Radiology 162:821–824, 1987.

Seibert, J.J., and James, C.A. (eds.): *Case Base Pediatric Radiology.* New York: Thieme, 1997.

Siegel, M.J., Glazer, H.S., St. Amour, T.E., and Rosenthal, D.D.: Lymphangiomas in children: MR imaging. Radiology 170:467–470, 1989.

Sinard, R.J., and Welling, D.B.: Cervical lymphangioma with simultaneous skull base invasion and soft tissue regression. Ann. Otol. Rhinol. Laryngol. 104:662–664, 1995.

Sumner, T.E., Volberg, F.M., Kiser, P.E., and Shaffner, L.D.: Mediastinal cystic hygroma in children. Pediatr. Radiol. 11:160–162, 1981.

Yakes, W.F.: Extremity venous malformations: Diagnosis and management. Semin. Intervent. Radiol. 11:332–339, 1994.

Yakes, W.F.: Diagnosis and management of venous malformations. In Savader, S.J., and Trerotola, S.O. (eds.): Venous Interventional Radiology with Clinical Perspectives. New York: Thieme, 1996, pp. 139–150.

Yakes, W.F., Luethke, J.M., Parker, S.H., et al.: Ethanol embolization of vascular malformations. RadioGraphics 10:787–796, 1990.

Yang, W.T., Ahuja, N., and Metreweli, C.: Sonographic features of head and neck hemangiomas and vascular malformations: Review of 23 patients. J. Ultrasound Med. 16:39–44, 1997.

Yoshida, H., Yusa, H., and Ueno, E.: Use of Doppler color flow imaging for differential diagnosis of vascular malformations: A preliminary report. J. Oral Maxillofac. Surg. 53:369–374, 1995.

Zadvinskis, D.P., Benson, M.T., Kerr, H.H., et al.: Congenital malformations of the cervicothoracic lymphatic system: Embryology and pathogenesis. RadioGraphics 12:1175–1189, 1992.

7

Psychosocial Impact of Congenital Vascular Lesions

JOSEPH BROGDON, M.A.

Facial disfiguration can cause significant psychopathological responses and behaviors in a patient and the patient's family. Facial disfigurement resulting from any physiological event or pathology is always tragic and produces tragic psychosocial outcomes. These outcomes are predictable and self evident in many patients with disfigurement, regardless of etiology. Studies regarding craniofacial defects and anomalies take many different experimental directions. The authors of this book are focusing specifically on the vascular lesioned or birthmarked child and have interwoven these studies with our own clinical experience, including diagnosis and treatment of hundreds of vascular lesions or malformations in children and adolescents.

In the usual childhood scenario, children have a family and/or live with a group of adults who function as a family system. When serious

Hemangiomas and Vascular Malformations of the Head and Neck, Edited by Waner, M.D. and Suen, M.D.
ISBN 0471-17597-8 © 1999 Wiley-Liss, Inc.

illness or disfigurement develops in any family member, some degree of psychosocial stress or suffering results. Such phenomena are even more painfully experienced when the afflicted is a small child. Critical questions include whether the impact is overwhelming to the family and child, how they respond to society's reaction to the disfigurement or how the family responds to the disfigurement. This dilemma is addressed in this chapter.

Answers are derived by the interweaving of the severity of the pathology, the variety of the symptomatology, the family system's functioning level, the resources of the child and family, spirituality, and, critically, how health care professionals understand and address the psychological effect on the child with a vascular lesion. Of critical importance is the promptness of professional intervention.

7.1. DEVELOPMENTAL OVERVIEW

Accurate, current information regarding the effects and dynamics of hemangiomas and other vascular lesion syndromes is a necessity for the practitioner treating children. Major events and stages of emotion and personality development are occurring from birth forward. Although babies are inarticulate about their inner life and even though they know virtually nothing about human relationships or concepts of self-esteem, they are often wide awake and alert, taking in and processing information. They are visibly perceptive as they make their first emerging contact with the world.

During the first year of life, a baby's emotional experience may range from the primitive to rapid acceleration into a much wider spectrum of feelings and perception. By 3–4 months of age, babies can respond consistently to events around them by smiling and even laughing. Reflections of sadness or disappointment also emerge about this time. Then, anger surfaces when the baby is denied its wants, and by 6–7 months of age, they begin anticipating visual events and can even show surprise if the expected does not occur.

Although the infant in its first year may not be intimately involved in emotional events occurring in their environment, their own affect and mood may be determined by the attitudes and moods they observe in their perceptual field, most notably their caretakers. This is manifestly evident in the first year or so of their life as they begin evolving a sense of self and self-awareness. They begin to sense the feelings of others, to show the rudiments of empathy.

This condensed road map of early development can serve as a template for better understanding how vascular lesions can and do impact the emotional development of the emerging self of a young child. As early intervention during the initial stages of life is so important, the physician needs to maintain the awareness that disfigurement impacts and affects the child, the child's family, peers, and even insignificant others, all of whom provide expressional and emotional feedback to the child. It is in these early formative months of life that innumerable developmental changes and milestones are encountered and reached by the child emerging into selfhood.

7.2. THE PSYCHOLOGY OF ATTRACTIVENESS

As early as the 1950s, numerous studies on attractiveness and its effect on social advantage, self-esteem, marital and vocational success, and so forth, began emerging.

7.2.1. The Attractiveness Stereotype: "What is Beautiful is Good"

In 1972, frequently cited, challenged and replicated perception studies (Dion and Bersheid, 1972) were published. The authors state that their study supports the hypothesis that a "what is beautiful is good" stereotype regarding physical attractiveness exists. Their data also support the hypotheses that

- Attractive persons are deemed to have more socially desirable personalities than unattractive individuals
- Attractive men and women are expected to achieve higher, more respected professional occupations than the unattractive
- Attractive people would be better marital spouses and have happier marriages
- The more attractive have a definite advantage in the dating scenario

· Pretty women are less likely to be thought of as guilty of a crime and are even more infrequently convicted of crimes than more unattractive women

The above study was replicated in 1975 by Dermer and Thiel to assess the generality of the "what is beautiful is good" stereotype. Cited in this study is that of Bersheid and Walster (1972), who reviewed a number of studies suggesting that "women's physical attractiveness levels covary positively with upward social mobility, as well as their self-reports of dating popularity." Consequently, the author implies that less attractive women are less capable of distorting their attractiveness positively and most apt to perceive themselves as inadequate and unsuccessful competitors with the more beautiful for the "rewards men may offer in their own society."

Although the purpose and scope of Dermer and Thiel's work had numerous other hypotheses and directions, for the purpose of this treatise, their study does support the contention that "attractive women are expected to be more sociable, heterosexually alluring, professionally successful and personally happy in comparison to unattractive women." Dermer and Thiel (1975) state that these factors are relevant and supported by their data and that a physical attractiveness stereotype does exist, but other variables have effect and the content of the stereotype is not always perfectly congruent with the "what is beautiful is good" thesis.

Regardless, the usual assumptions that beautiful is better and that more attractive people become more successful in all areas are borne out by perceptual studies in this area. However, the adage that "beauty is in the eye of the beholder" is relevant. Such variables as the attractiveness of the beholder clearly biases the perception of the beheld's attractiveness.

7.2.2. The Advantages of Attractiveness

Even more relevant to this book's direction regarding the need for early intervention and treatment of hemangiomas, Clifford et al. (1973) explored how a child's physical attractiveness biased elementary schoolteacher expectations. Investigating further, the hypothesis of Rosenthal and Jacobson (1968) "that a teacher's expectations as to how a child *will* behave have an enormous impact on how the child *does* behave." Numerous other studies are cited, demonstrating that

- First impressions affect subsequent interactions
- Expectations do influence behavior
- Positive relationships exist between teacher attitudes and the students' classroom performance

Clifford et al. (1973) postulate that the two most important and usual sources of information about a student from which a teacher may develop an initial bias are the child's school record and the child's general appearance. Clifford et al. examined the impact of the latter variable while holding the former constant. Fifth grade teachers were the subjects of the study and were provided data on 441 fifth grade students. Specifically, they were provided the student's school record and the student's photograph and asked to complete an opinion sheet regarding their estimates of

- The child's probable I.Q.
- The child's peer relationships
- The child's parents' attitude toward school
- Prediction of the child's level of schooling (e.g., high school up to doctorate level)

The hypothesis of this study was confirmed. As predicted, attractive children clearly maintained a sizable advantage over the less attractive children. The teachers perceived attractive children to have a higher educational potential, to have a superior I.Q., and to have educationally involved parents. Teachers also expected attractive children to have more positive and successful peer relationships and to be better liked by peers than less attractive children. Furthermore, the gender of neither the child being rated nor the teacher rating the child significantly affected the outcome of the ratings.

Clifford et al. (1975) conclude that educators and parents alike should be sensitive to the tremendous impact the child's attractiveness or lack of attractiveness might have on the ways the child is perceived and treated by others, peers and authority figures alike, and, further-more, they speculate that many variables affecting a child's attractive-ness can be manipulated and assisted and that all concerned parties should be aware of how these variables can "operate as an unwar-ranted detriment to his (her) intellectual development."

7.2.3. Psychopathology and Unattractiveness

The clarity with which McCabe and Marwit (1993) explore the relation-ship between a child's self-concept and the likelihood of developing depressive symptomatology cogently makes the case for early treat-ment of craniofacial anomalies, specifically, vascular lesions. This study reveals significant correlations.

McCabe et al. also mention other studies (e.g., Abbot and Sebastian, 1981) supporting the contention that self-perception of attractiveness alone has more psychosocial effect on the child than more objective rat-ings of others in their social context, even in expectations of success. Additional studies cited (Lerner et al., 1973; Mendelsohn et al., 1981; Secord et al., 1982; Starr, 1982; Rohrbacher, 1973; Rosen et al., 1968) also emphasize the importance of self-rated attractiveness and its correla-tion with positive self-esteem and self-concept.

Beck (1967) advanced the concept that physical attractiveness and body image are highly correlated with the absence of presence of dys-phoric or affective symptomatology.

McCabe et al. (1993), utilizing multiple regression analyses, revealed that "dissatisfaction with the body's attractiveness was the most pow-erful single variable in predicting dysphoria": $F(1, 55) = 6.54$; $p < 0.02$. This study supported the following:

- Negative self-evaluation of attractiveness is related to affective disorders
- Depressed children are most specifically unhappy with their per-ception of body appearance and less impacted by others' percep-tions
- Depressed children negatively distort their own attractiveness, while remaining more accurate about others' attractiveness

This study exemplifies the importance of the relationship between the parents and affective symptoms.

All these data are convincing and relevant to our purpose here—supporting the belief in the need for early intervention in treating craniofacial anomalies in children. It is clear that as children with hemangiomas, for example, mature and grow, they are more likely to have low self-esteem and negative body image. Evident also, attractiveness somehow has exceptional impact on social and vocational success and evaluation of success potential in these important life factors and even possible intellectual attainment. It is further clear that children with hemangiomas can be treated, that their attractiveness can be changed or upgraded, and that this intervention is very important to the child's educational, social, psychological, emotional, and intellectual development.

7.3. PSYCHOSOCIAL SCOPE OF THE PHENOMENON

There are now numerous comprehensive, well-designed and well-controlled studies on the psychosocial effects of vascular lesions in young children. Until recently, relevant studies were few. This lack seemed primarily due to the fact that the traditional biomedical diagnostic and treatment approach to vascular lesions has been one of nonintervention or benign neglect. This strategy frequently results in unintended, but tragic, disfiguring consequences. This policy is now being challenged by aggressive, early, and successful intervention.

Sheerin et al. (1995) state that "facial birthmarks have the most devastating effects" in young children in that the more severe the disfigurement, the more profound the impact on the child, the child's developmental life, and the child's self-concept. Pragmatically evident is that the more advanced the proliferation, the more devastating the psychological and emotional consequences without surgical intervention. Sheerin et al. (1995) indicate that, although numerous studies have been quoted as evidence for this scenario, most are anecdotal (Bull, 1990; Landsdown et al., 1991). They conclude, however, that "psychosocial adjustment varied according to the nature of the disfigurement or deformity and was unrelated to the severity of the disfigurement."

Some of the studies in this area failed to use reliable measurements and control groups, or relevant demographic normative data were absent. Considering the population being addressed, however, this is not

surprising. Many related studies in this context can provide useful in-
put and impetus for further research with solid experimental design
being implemented.

A related study by Hoare et al. (1993) utilized the Self-Perception
Profile For Children (Harter, 1985), a multidimensional developmental
perspective of a child's self, which was administered to 5,000 Scottish
children. part of Hoare's findings was that the Physical Appearance
subscale was significantly correlated with overall self worth. This is described as supporting Hoare's claim that "self-perceptions about physical appearance invariably account for the major proportion of the vari-
ance of feelings of general self worth or self esteem" (Lerner and
Jovanovic, 1990).

7.4. CURRENT STATE-OF-THE-ART
CLINICAL RESPONSE

This author is gratified to have the opportunity to interact with, inter-
view and evaluate the families of birthmarked children, including the
birthmarked child, when possible. Uniqueness and specialness are
commonly encountered in these families.

Many different clinical attitudes, perspectives, and approaches to
treatment of congenital vascular lesions have been and are being encountered by the parents and families of the vascular-lesioned child.
Simply stated, there is a preponderance of denial present in the medical
practitioners treating these families, denial in that many of these practitioners are too conservative or fearful of aggressive, appropriate surgical intervention. This response (or lack of response) frequently results
in additional psychosocial trauma to both the child and family. This
"wait and see" conservative attitude can and does result in facial disfigurement in many children.

Data indicate that even those children whose hemangiomas involute rapidly (before 6 years of age), 40% still require corrective surgery.
In children whose lesions involute more slowly, the projections are
even more dismal. Eighty percent of children with slow involuting hemangiomas require some form of corrective surgery (Finn et al., 1983).

Unfortunately, the number of unsuccessfully treated (or untreated)
children who have been managed using this clinical approach is unknown at present. A research project investigating this patient popula-

tion is being considered currently. We at Arkansas Children's Hospital only see the vascular-lesioned child who is fortunate to have caring parents who are consistently aggressive in seeking out the most viable and effective treatment available.

7.5. PSYCHOLOGICAL SEQUELA OF THE VASCULAR-LESIONED CHILD

Definitive exploration of the psychological effects has proved to be useful and convincing regarding the call for early intervention. Numerous investigations have evaluated the psychological sequelae of craniofacial anomalies (Sheerin et al., 1995; Corah and Corah, 1963; Lanigan and Cotterill, 1989; Goldstein, 1993; Nordlicht, 1979).

Although some of these studies did not examine specifically the psychological effects of vascular lesions, the studies investigated are all categorized under the more generic heading of craniofacial anomalies. These disfigurement studies, while not exact, are perceived as relevant. A facially disfigured child is a facially disfigured child, and the disfigurement speaks for itself.

Personality profiles of adolescents with dissimilar disabilities (cleft lip/palate, orthopedic) were evaluated by Harper et al. (1978). Even though the subjects rated in this study were not birthmarked children, we take the liberty of extrapolating from the data of this sample to the birthmarked child with confidence.

Harper et al. (1978) interpret the findings to support the hypothesis that disfigurement has a significant psychological impact on the child, often in different ways. Using the Minnesota Multiphasic Personality Inventory (MMPI), Harper et al. suggest that adolescents with impairment from cleft palate display greater self-concern and ruminative self-doubt in interpersonal interactions. The orthopedically impaired group in this study presents isolated and passive orientation in interpersonal relationships. Harper et al. (1978) also suggest that their subjects' MMPI profiles reflect that the fact of impairment, rather than the type, is of more significance in examining the effect on psychosocial behavior patterns. These findings also suggest that mildly impaired adolescents from both groups presented with a significantly greater degree of inhibition socially than matched normals. Further indicated is that mildly impaired (disfigured) adolescents from both groups show more social inhibition than severely impaired adolescents. This trend is reported in other studies also.

An excellent examination of trends and issues in studying children with significant differences is that of Sigelman et al. (1986) in the text-book *The Dilemma of Difference*. The editors of this volume call for more extensive and controlled studies of the stigmatizing attributes and how they affect children developmentally, in which more attention should be directed to individual differences and stigmatizing reactions. Sigelman et al. address this issue from various angles of differences in developmental theory (psychoanalytic versus social learning versus cognitive-developmental). They delve into whether this tendency to stigmatize others is based on differences and suggests that even the potential for stigmatization lies in everyone and that the development potential "may come to depend less on cognitive universals and more on social experience as a child moves out of infancy and into childhood." Several researchers' results indicate that young children exhibit preference for the "able-bodied over the nonable-bodied." Other studies of the behavioral component of attitude also suggest preference for those without disabilities. Furthermore, these authors suggest that as children move from preschool to school age years, negative attitudes toward disabled children become more positive with age, just as certain aspects of racial prejudice do. As early as 1951, MacGregor noted that the facially disfigured frequently encountered barriers to the privileges and opportunities available to the nonhandicapped person. Such an affliction is more a social than a physical one. The stigmatization results from the visibility of the defect, and the degree of reaction is due in large part to the profound social significance of the face, which is generally the center of attention in any human interaction. "If he looks different, he must, ipso facto, be different." this differentness results in biases and misunderstandings that equate with those encountered in minority and majority group relationships. MacGregor (1951) suggests that not only do such defects tend to operate to the individual's social and economic disadvantage, but that these disadvantages and attitudes tend to determine the facially disfigured person's self-image, attitude toward himself, and his own mental health. Furthermore, the author states that "wherever plastic surgery can correct or improve the facial injury or congenital malformation, it should be undertaken as early as possible in order to avoid not only the obvious disadvantages, but to prevent deep psychological wounds that may be incurred, but not so easily limited." This is a very interesting and precocious clinical perspective, considering it was proposed over 46 years ago.

Another interesting study by Kalik and colleagues (1981) explored the social issues of body image concerns of port-wine–stained patients undergoing different stages of laser treatment (ages ranged from 7 to 66 years, with a median of 24 years). They point out two important features of the port-wine–stained patient that distinguish them from the hemangioma patients: Because the port-wine stain itself does not affect configuration or shape of the face, the port wine stain is readily perceived as an element foreign from one's face; and the severity of distress expressed by these patients is sometimes out of proportion to the degree of disfigurement. Significant here is that port-wine stain marks bear as "different" rather than simply "unattractive." These patients were administered several psychological test instruments (Eysenck Personality Inventory, Multiple Affect Adjective Checklist, and the State-Trait Anxiety Inventory). In summary, the authors indicate that, as a group, the patients did not exhibit an unusual degree of emotional disturbance compared with the normal, control population, but that their initial expectations of laser therapy tended to be "unrealistically perfectionistic." The authors conclude that, among their sample, the port-wine stain had been a source of stress and psychological burden. Of interest is that this particular subgroup of facial disfigurement patients is deemed by these authors as having managed to maintain a healthy self-concept and not to have developed significant psychopathology as a result of the disfigurement. The authors further stress that any physician treating any kind of facial anomaly, specifically, port-wine stains, needs to be well aware of and appreciate the psychological and social aspects of body image, as well as the physical aspect.

Pillemar and Cook (1989), studying the psychosocial adjustments of pediatric craniofacial patients after surgery, concluded that the older the children were at the time of surgery, the poorer the self-concepts they had and the more depressed and withdrawn they were. These authors also stated that children with craniofacial anomalies are more likely to evince an inhibited personality style, low self-esteem, impaired peer relationships, and greater dependence on caretakers compared with a control group of normative samples. This study also explored teacher–student relationships in these patients and found underachievement in the patient population due to overdependence on teachers and to limited peer support both academically and socially.

Even more significant, Pillemar and Cook (1989) stated that children who were more attractive postsurgically (as determined by independent rating of photographs) tended to have better self-concepts.
Data here even suggest that the stress induced by facial disfiguration
impacts the family and that parents who were more highly stressed by
life events rated their children as more depressed, withdrawn, and socially inept. The authors also conclude that corrective surgery in and of
itself is not a panacea. Timely and rapid intervention interfaced with
integration of biomedical and psychological expertise ensures even
more positive outcomes in offering therapeutic support and aftercare
following craniofacial surgery.

A study by Pertschuk and Whitaker (1982) indicates that "while
surgeons describe a number of indications for craniofacial surgery, psychological benefit is usually at or near the top of the list" and cites studies supporting this position (Edgerton, et al., 1974; Marsh, 1980;
Whitaker and Randall, 1974). They further agree with Pertschuk's concepts covered above regarding social and psychological effects of attractiveness versus unattractiveness.

Perkschuk's interdisciplinary surgical team utilized pre- and postsurgical evaluation protocols, including interviews with patients and
family, and psychological and psychosocial assessments of the patients
when appropriate. No control groups were utilized, however. Their
conclusions include that the younger the surgical patient and the earlier the intervention, the more positive the outcomes.

The interpretation made is that the older children are more likely to
present as more socially isolated, have more self-concept problems, and
to be more subject to affective difficulties. Critical is that Pertschuk
and Whitaker (1982) infer significant correlations between appearance
improvement and positive changes in self-concept and anxiety.
Improvement in psychosocial adjustment accompanies craniofacial surgical alteration, resulting in upgraded facial attractiveness.

Harrison's treatise (1988) provides one approach in assisting the
physician faced with dilemmas inherent in diagnosis and treatment of
birthmarks. She suggests using a symptom hierarchy in sorting out the
effects of a birthmark on the child's emotional state. Harrison proposes
using the "best case–worst case" approach rather than specific guide-
lines for decision-making.

Harrison views a birthmark in a child of any age to be a significant
life stressor and having a stage-specific impact on both child and fam-

ily, another reference that clearly supports the need for early intervention with vascular lesions.

Harrison (1988) also emphasizes that when parents see beyond the disfigurement, sees the child as a complicated person in his or her own right, they often deny or overlook the birthmark until others call their attention to it. "I forget about the birthmark until I see a stranger's reaction to it. To me, she is just my Nancy, and she is beautiful."

7.6. THE FAMILY

In our clinical experience with parents at Arkansas Children's Hospital, we repeatedly encounter unconditional expression of love for their birthmarked child. As Harrison (1988) indicates, these caring parents of birthmarked children inspire us all.

A very touching, personalized narrative in *Good Housekeeping* (Goldstein, 1993) written by the mother of a birthmarked child, captures the agony, uncertainties, and unconditional acceptance parents experience in their treatment journey. This story mirrors a typical scenario presented by the families interviewed and treated at Arkansas Children's Hospital. Described is a barely noticeable red mark or "scratch" evident at birth, which in 6 weeks rapidly proliferated from a red mark to a growth extending from the child's left eye to her chin, completely blocking her left nostril, impairing respiration, and distorting her mouth to such a degree that she could not retain fluids for feeding. The pediatric plastic surgeon assured the parents that invasive surgical procedures were inadvisable, that the hemangioma would undergo involution, and he recommended waiting until the child was 3–4 years old before initiating surgical correction. Complications included the dangers of loss of eyesight, as the growth had begun occluding a large percentage of her vision.

The mother also describes the pain of being subjected to the frequent insensitivity and intrusion of even strangers in public places in response to her child's disfigurement. The resultant anger and resentment so often seen in Arkansas Children's Hospital patients are poignantly depicted by this child's mother. Typical confrontations by the mother occurred, and she eventually became more isolated socially and more embedded in defensive denial. Even more compelling is how the child at age 2 years "discovered" her disfigurement, and in the process her mother felt helpless to protect the child from her own self-awareness of disfiguration. Then, at age 2 and a half, this child finally underwent her first corrective surgical procedure, then another at 4, then another one a year later, which resulted in 75% removal of the lesion. The family was told that additional surgery on the remaining 25% disfiguration could not be done until the child was 9–10 years old. In conclusion, this mother states (Goldstein, 1993):

> This story is not my story and in many ways, it is not Morgan's story. It is the story about all the children who do not easily blend with the images of beauty we have created or the standards of acceptability we have set. It is the story of the countless numbers of parents who silently endure the pain of raising a child with a disfigurement or disability. The next time you see our children in the street, be thankful for all you have—but don't pity us. We have been given the same gift as you, just look a little deeper.

REFERENCES

Abbot, A. and Sebastian, R.: Physical attractiveness and expectations of success. Pers. Soc. Psy. Bull. 7:481–486, 1981.

Beck, A: *Depression: Clinical, Experimental and Theoretical Aspects.* New York; Hoebor, 1967.

Berscheid, E., and Walster, E.: Beauty and the best. Psy. Today 5:42–46; 74. 1972.

Bull, R.N.: Society's reaction to facial disfigurement. Dental Update PP202–205, Special Update/June 1990.

Clifford, M. and Walster, E.: The effect of physical attractiveness on teacher expectations. Soc. Ed. 46:248–258, 1973.

Dermer, M., and Thiel, D.: When beauty may fail. J. Pers. Soc. Psychol. 31:1168–1176, 1975.

Dion, K.: Physical attractiveness and evaluations of children's transgressions. J. Pers. Soc. Psychol. 24:207–213, 1972.

Dion, K., and Bersheid, E.: What is beautiful is good. J. Pers. Soc. Psychol. 24:285–290, 1972.

Edgerton, M., Jane, J., and Berry, F.: Craniofacial osteotomies and reconstructions in infants and young children. Plast. Reconstr. Surg. 54:13, 1974.

Finn, M., Colowacki, J., and Milliken, J.B.: Congenital vascular lesions: Clinical approach of a new classification. J. Pediatr. Surg. 18:894, 1983.

Goldstein, R.: Making things right for Morgan. Good Housekeeping 216(6):54, 1993.

Harrison, A.: The emotional impact of a vascular birthmark. In *Vascular Birthmarks Hemangiomas and Malformations,* Mulliken J. and Young A.P. (eds.) 454–462, Philadelphia: W. B. Saunders Company, 1988.

Harter, J.: Manual for the Self-Perception Profile for Children. Denver: University of Denver, 1985.

Harper, D., and Richman, L.: Personality profiles of physically impaired adolescents. J. Clin. Psychol. 34:636–642, 1978.

Hoare, P., Elton, R., Greer, A., and Kerley, S.: The standardization of the Harter questionnaire with Scottish children. Eur. Child Adolesc. Psychiatry, 1993.

Kalik, S., Goldwyn, R., and Noe, J.: Social issues and body image concerns of port wine stain patients undergoing laser therapy. Lasers Surg. Med. 1:205–213, 1981.

Lanigan, S., and Cotterill J.: Psychological disabilities amongst patients with port wine stains. Br. J. Dermatol. 121:209–215, 1989.

Lansdown, R., Lloyd, J., and Hunter, J.: Facial deformity in childhood: Severity and psychological adjustment. Child Care Health Dev. 18:165–171, 1991.

Lerner, R.M., and Jovanovic, J.: The role of body image in psychosocial development across the lifespan: A developmental contextual perspective. In Cash, T.F., and Pruzinsky, T. (eds.): *Body Images: Developmental Deviance and Change.* pps. 110–130, New York: Guilford Press, 1990.

Lerner, R., Orlos, J., and Knapp, J.: Physical attractiveness, physical effectiveness and self-concept in late adolescents. Adolescence 11:313–326, 1976.

MacGregor, F.: Some psychosocial problems associated with facial deformities. Am. Soc. Rev. 16:629–638, 1951.

Marsh, J.: Comprehensive care for craniofacial anomalies. Curr. Probl. Pediatr. 10:1, 1980.

McCabe, M., and Marwit S.J.: Depressive symptomology, perceptions of attractiveness, and body image in children. J. Child. Psy. 34(7):1117–1124, 1993.

Mendelson, B., and White, D.: Relation between body-esteem and self-esteem of obese and normal children. Perceptual Motor Skills 54:899–905, 1982.

Nordlicht, S.: Facial disfigurement and psychiatric sequelae. N.Y. State J. Med. 1382–1384, 1979.

Pertschuk, M., and Whitaker, L.: Social and psychological effects of craniofacial deformity and surgical reconstruction. Symposium on Social and Psychgological Considerations. Clin. Plast. Surg. 9:287–306, 1982.

Pillemar, F., and Cook, K. (eds.): The psychosocial adjustment of pediatric craniofacial patients after surgery. Cleft Palate J. 26:201–207, 1989.

Rohrbacher, R.: Influence of a special camp for obese boys on weight loss, self-concept, and body image. Res. Q. 44:150–157, 1973.

Rosen, G., and Ross A.: Relationships of body image to self-concept. J. Cons. Clin. Psy. 32:100, 1968.

Rosenthal, R., and Jacobson, L.: *Pygmalion in the Classroom.* New York: Holt, Rinehart and Winston, 1968.

Secord, P., and Jourard, S.: The appraisal of body-cathexis and the self. J. Cons. Clin. Psy. 17:343–347, 1953.

Sheerin, D., Macleod, M., and Kusumakar, V.: Psychosocial adjustment in children with port wine stains and prominent ears. J. Am. Acad. Child Adolesc. Psychiatry 34:1637, 1995.

Sigelman, et al.: Stigmatization in childhood. In Ainley, S., Becker, G., Coleman, L. (eds.): The Dilemma of Difference, Ch. 10, pp. 185–208. New York and London: Plenum Press, 1978.

Starr, P.: Physical attractiveness and self-esteem ratings of young adults with cleft lip and/or palate. Psy. Rep. 50:467–470, 1982.

Whitaker, I., and Randall, P.: The developing field of craniofacial surgery. Pediatrics 54:574, 1974.

Treatment Options for the Management of Hemangiomas

MILTON WANER, M.D., F.C.S. (S.A.) AND JAMES Y. SUEN, M.D., F.A.C.S.

Until recently, the policy of "benign neglect" was so firmly entrenched that even to consider treating a noncomplicated hemangioma was pure heresy. To all but a few patients with life- or sight-threatening lesions, there were no options. Despite this, several physicians have realized the need for an alternative and have developed an expertise in the means that fell within the domain of their specialty. Those with surgical training may be more inclined to favor a surgical alternative, whereas to an interventional radiologist embolization may seem the logical approach. A dermatologist familiar with the use of a laser may use this modality most often, whereas one unfamiliar with lasers may favor pharmacotherapy. It is therefore patently obvious that several treatment modalities may be used and that each of these probably has a

Hemangiomas and Vascular Malformations of the Head and Neck, Edited by Waner, M.D. and Suen, M.D.
ISBN 0471-17597-8 © 1999 Wiley-Liss, Inc.

legitimate role in the management of hemangiomas. Unfortunately, because active intervention is only just moving into mainstream medicine, we have all tended to work in a vacuum and therefore no controlled clinical trails comparing these modalities have been undertaken. In the best interests of our patients, it is imperative that we establish protocols of management based on what is best for the patient rather than what we are most familiar with. A review of all of the commonly used modalities is thus of paramount importance.

Presently, all of the treatment modalities can be considered under 4 main categories: pharmacotherapy, laser treatment, surgical excision, and interventional radiology.

8.1. PHARMACOTHERAPY

This category includes the use of any pharmacological agent. Steroids and interferon are the most commonly used, and, although other agents are known to modulate angiogenesis, none have been advocated.

8.1.1. Steroids

The first reports of the effects of steroids on hemangiomas appeared in the mid to late 1960s (Katz, 1965; Zarem and Edgerton, 1967; Fost and Esterly, 1968). Since then, prednisone and prednisolone have become the standard of care as first-line drugs for life- or sight-threatening hemangiomas (Edgerton, 1976; Enjolras et al., 1990). The mechanism of action of steroids, however, remains unknown. Early experimental work suggested that steroids produced vasoconstriction of arterioles and precapillaries (Zweifach et al., 1953; Wyman et al., 1953). More recent work, however, has shown that this effect is indirect. Glucocorticoids potentiate the effect of epinephrine and norepinephrine on vascular smooth muscle (Wilfingseder and Propst, 1972; Hiles and Pilchard, 1971). The effect of synthetic glucocorticocoid drugs on proliferating hemangiomas is thus likely to be similar. Prednisone and prednisolone probably potentiate the effect of certain vasoconstrictors on vascular smooth muscle, and this may at least in part account for their clinical inhibition of proliferating hemangiomas. Folkman and his coworkers found that hydrocortisone given together with a constituent of heparin prevented angiogenesis in experimental tumors (Crum et al., 1985). The same effect was observed only with very high doses of methylprednisolone and was not seen with dexamethasone. Clinical experience with dexamethasone, is not however, consistent with these findings. Dexamethasone has an extremely potent inhibitory effect on proliferating hemangiomas. Sasaki et al. (1984) reported the presence of estradiol receptors in hemangiomas. This may account for the higher incidence of hemangiomas in females. Steroids appeared to block these receptors in vitro. Unfortunately, the presence of these receptors in "cavernous hemangiomas" was not demonstrated. However, their exact definition of cavernous hemangioma is not clear. Sasaki et al. (1984) could possibly have used a venous malformation that in those days was still regarded as a "cavernous hemangioma."

Reports of the response rates to steroids vary from 30% (Bartoshesky et al., 1978; Enjolras et al., 1990) to 93% (Sadan and Wolach, 1996). Three factors appear to be responsible for this great variation:

1. The variety of dose regimens: Earlier investigators, namely, Edgerton (1976), Fost and Esterly (1968), and Brown et al. (1972), used 20–40 mg/day regardless of the child's weight, whereas most practitioners today will titrate the dose to the body weight. Enjolras et al. (1990) and Mulliken et al. (1995) recommend 2–3 mg/kg body weight, and Sadan and Wolach (1996) use 5 mg/kg body weight. Not surprisingly, physicians who used a higher dose reported better response rates.

2. The duration of treatment: Clearly, some investigators kept their patients on steroids for much longer than others. Once again, those who used steroids for longer periods reported greater success, with the exception of Enjolras et al. (1990), who, despite longer regimens, reported poorer results. However, they did use a lower dose (2–3 mg/kg).

3. The age of onset of treatment: Because steroids only appear to be effective during proliferation, patients in whom proliferation has ceased will not respond (Fost and Esterly, 1968; Cohen and Wang, 1972; Bartoshesky et al., 1978). Sadan and Wolach (1996), who reported a 93% success rate, commenced treatment before 6 months of age in 53 of their 60 patients. Fost and Esterly (1968) and Cohen

and Wang (1972) also reported a better response in younger children, whereas Enjolras et al. (1990) reported no relationship between the response rate and age of initiation of treatment.

Potential side effects include cushingoid features (moon facies, buffalo hump, and central obesity), growth retardation, gastroesophageal reflux, peptic irritation and ulceration, fluid and electrolyte disturbances, hypertension, hyperglycemia and glycosuria, behavioral disturbances, anorexia, and immune suppression (Fig. 8.1). With regard to immune suppression, an often cited study, reported by Gunn et al. (1981), showed a decrease in the number of T lymphocytes and an increase in the number of childhood infections in children who received

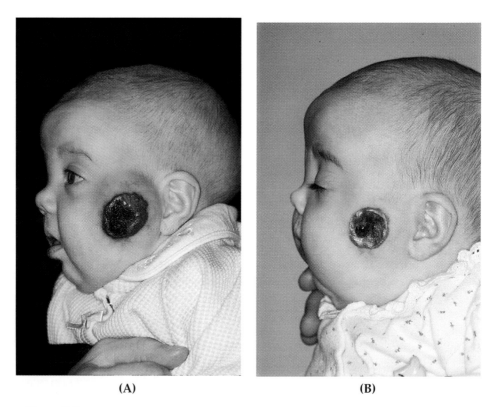

(A)   (B)

Figure 8.1
(A, B) An infant before and 6 weeks after treatment with Prednisolone (4 mg/kg). Note the reduction in the volume of the hemangioma. The ulcer is still present, and cushingoid features have appeared.

only two doses of steroids for possible respiratory distress syndrome. No difference in the white cell count, immunoglobulin levels, or the complement levels were found. A major criticism of this study concerns the fact that only 14 children were studied. The significance of this study thus remains unclear. In spite of the considerable list of potential side effects, most reports claim few side effects and no long-term effects (Zarem and Edgerton, 1967; Katz and Askin, 1968; Fost and Esterly, 1968; Cohen and Wang, 1972; Edgerton, 1976; Sadan and Wolach, 1996). More importantly, no reports of opportunistic infections were found. We can thus assume that this is an extremely rare phenomenon. Sadan and Wolach (1996), who used the highest dosage regimen and administered steroids for the longest period of time, reported a 50% incidence of cushingoid features, and only 2 of 60 patients experienced growth retardation. Changes in behavior were noted in almost all their patients: irritability, frequent crying, and increased appetite. All of these side effects reversed soon after the cessation of treatment (Sadan and Wolach, 1996).

Prednisone and prednisolone, by virtue of their effect on angiogenesis, will thus only be useful during proliferation and not during involution. Steroids should therefore only be used during the first year of life and, better still, during the first 6–8 months of life (the period of rapid proliferation). As proliferation slows during the latter part of the first year, steroids will become less effective. Sufficiently high doses should be used and for a long enough period of time. We recommend the following regimen:

- Prednisone or prednisolone, 5 mg/kg body weight, should be administered as a once a day morning dose.

- Appropriate doses of Zantac and Propulsid should be given simultaneously to prevent gastritis and gastroesophageal reflux.

- If no response is seen within 1 week, the dosage (of prednisone or prednisolone) should be discontinued.

- If a response is seen, in the form of a diminution in the size of the lesion or a cessation in the growth of the lesion, the full dosage should be continued for at least 2–3 weeks.

- Following this, the dose should be gradually tapered down over an 8–10 week period.

· If during or after the taper, rebound proliferation is noted, the dosage should once again be increased to the next higher dose and maintained at this level for a further week after which a tapering regimen should once again be resumed.

· A repeat of rebound proliferation should, as before, prompt the physician to increase the dosage for a further 2 weeks and then again attempt a tapering dose.

8.1.1.1. Intralesional Steroids

Intralesional injections of steroids were first introduced in an attempt to overcome the steroidal systemic effects (Azzolini and Nouvenne, 1970). This method has since become established as an appropriate modality for the treatment of well-localized hemangiomas, such as periorbital lesions (Kushner, 1979, 1985; Zak and Morin, 1981; Brown and Huffaker, 1982). Kushner (1979, 1985) recommends using a combination of triamcinolone and betamethasone. His rationale for this is that combining the short-acting betamethasone with the long-acting triamcinolone will ensure an immediate effect because the effect of triamcinolone will not be seen for several days. The recommended dosages are 3–5 mg/kg body weight of triamcinolone and 0.5–1.0 mg/kg body weight of betamethasone. The two agents should be injected separately with separate syringes. The steroids should be injected into the substance of the lesion with 1-ml syringes and 27-guage needles. In the presence of a large lesion, an attempt should be made to distribute the substance throughout the tumor. This may require several passes. It may be necessary to perform the procedure under general anesthesia, especially when dealing with a periorbital hemangioma.

As is the case with systemic steroids, response rates vary and appear to be similar to those with systemic steroids (Kushner, 1985; Sloan et al., 1989). Furthermore, the effect of a single intralesional injection is short lived and may need to be repeated at least once 6 weeks later (Kushner, 1985).

Complications from local injections include retinal artery occlusion of both the ipsilateral eye (Shorr and Seiff, 1986) and the contralateral eye (Ruttum et al., 1993), eyelid necrosis (Sutula and Glover, 1987), and atrophic scarring (Droste et al., 1988). In addition, systemic effects of adrenal suppression and growth retardation have been reported (Weiss, 1989). The mechanism of several of these side effects stems from the fact that the high-flow system of vessels that make up the hemangioma is continuous with the systemic and, in the case of a periorbital lesion, the orbital circulation. The injected material may thus embolize into the retinal arteries and occlude them or into the venous system in which case systemic effects can be anticipated. For all these reasons, some physicians are reluctant to inject periocular hemangiomas with intralesional steroids (Edgerton, 1976; Deans et al., 1992). While we acknowledge that the likelihood of serious complications is low, their severity must be considered when treating periocular hemangiomas. We thus rarely use intralesional steroids.

8.1.1.2. *Topical Steroids*

In yet another attempt to overcome the systemic effects of steroids and also to avoid the serious risks of intralesional injections, Elsas and Lewis (1994) treated five patients with topically applied clobetasol, a high-potency steroid. While they were able to show a response in all cases, the effect was not as dramatic as one might expect with both systemic and intralesional steroids (Elsas and Lewis, 1994). Topical clobetasol was unable to relieve astigmatism, and, as is the case with other methods of steroid administration, prolonged periods of administration are necessary. Furthermore, cases of adrenocortical suppression with topical clobetasol have been reported. It seems likely therefore that the mechanisms of action of both topical and intralesional steroids are at least in part systemic. Without substantiated evidence of either a greater efficacy or a reduction in the number of side effects, topical and/or intralesional steroids should not replace the systemic route of administration as the most frequent route.

8.1.2. Interferon

Interferon was originally used as an antiviral agent but was subsequently found to inhibit endothelial cell migration and proliferation as well as several other steps in angiogenesis (Brouty-Boyce and Zetter, 1980; Friesel et al., 1987; Feldman et al., 1988; Folkman, 1989; Tsuroka et al., 1988). Interferon will also inhibit the effects of specific growth factors such as endothelial cell growth factor or fibroblast growth factor (Heyns et al., 1985; Oleszak and Inglot, 1980). While being tested as a antiviral agent in a clinical trial with human immunodeficiency virus–infected patients, interferon was serendipitously found to have an effect on Kaposi's sarcoma (Groopman et al., 1984; Real et al., 1986; Rios et al., 1985). Since then, anecdotal reports of successful clinical outcomes have appeared in the literature (Orchard, 1989; Loughnan et al., 1992). Ezekowitz et al., (1992), Ohlms et al., (1994), and, more recently, Soumekh et al. (1996) published a prospective series in which a total of 50 patients diagnosed with hemangiomas were treated. Their results were fairly consistent. All reported accelerated regression of the lesion in the vast majority of their patients despite the fact that many of the patients had failed steroids and in several the lesion had stopped proliferating by the time treatment had commenced.

From this we can deduce that, unlike steroids, interferon does not have to be given during the proliferative phase to be effective. Furthermore, because the mechanism of action of interferon appears to differ from that of steroids, interferon may work despite a failed attempt at steroid therapy. Unfortunately, regression of the hemangioma appears to be much slower than with steroids. The dosage and the length of treatment used by all three groups of investigators is fairly consistent. Ezekowitz et al. (1992) used up to 3×10^6 units/m^2 of body surface area of interferon-α2a given as a daily subcutaneous injection, and their average duration of treatment was 7.8 months. Ohlms et al. (1994) treated patients with airway obstruction and used the same regimen except that their mean duration of treatment was slightly longer (11.3 months). Soumekh et al. (1996), on the other hand, used the same dose of interferon-α2b. Furthermore, unlike the previous two regimens, Soumekh et al. administered the interferon as a daily subcutaneous injection for the first 6 months and then three times per week for the next 6 months.

Toxicity was mild and transient. All patients experienced low-grade febrile reactions, especially at the beginning of treatment. Transient neutropenia, anemia, and a mild elevation in liver enzymes were also reported, and in one case a vigorous initial response with dermal necrosis of the involved skin was seen. Mulliken et al. (1995) reported up to a fivefold increase in liver transaminases. Because of these initial reports, a disturbing pattern appears to have developed. Vesikari et al. (1988) reported a case of a child with laryngeal papilloma who had been treated with human leukocyte interferon and who developed irreversible spastic diplegia. The child had been treated for a total of 11 months with doses varying from 2 million units every second day to 3 million units per day, and the symptoms developed while on the highest dose. After cessation of the interferon, the symptoms of spasticity improved but did not reverse. Several other cases of irreversible spastic diplegia have, since then, been reported in patients treated for hemangiomas (Mulliken et al., 1995).

Other side effects of interferon have been reported, most of which appear to be reversible. These have all been reported in patients treated for viral and neoplastic disorders and include fatigue, progressive mental and motor slowing, speech difficulties, and hallucinations. In addition to these, Guyer and coworkers (1993) reported the coincidental finding of retinal cotton-wool spots in 10 patients treated for choroidal neovascularization as well as various other disorders. In all but one patient no visual loss resulted, and the effects appeared to be reversible. Retinal cotton-wool spots result from ischemia, which Guyer et al. (1993) speculate is the result of immune complex deposition within the retinal vasculature. Leukocyte infiltration and subsequent retinal ischemia will then result (Guyer et al., 1993).

In view of these side effects, we recommend the use of interferon only for life- or sight-threatening complications and, with the exception of Kasabach-Merritt syndrome, only after steroids have been tried. The usual regimen is

- Baseline and monthly developmental and neurological evaluation
- Baseline and monthly complete blood count, liver transaminase, thyroid function studies, and urinary basic fibroblast growth factor levels
- 3×10^6 units/m^2 body surface area interferon-α2a given as a daily subcutaneous injection
- Patients seen monthly (if possible)
- Developmental and neurological examination 3 months after completion of therapy

This regimen should be continued as long as there is a response and until the lesion is no longer life threatening and/or the reason for prescribing interferon no longer exists. This may be as long as 6–8 months. If, on the other hand, the lesion fails to respond after a reasonable trial of at least 1 month, interferon should be discontinued. Unlike steroids, there is no hypothalamic-pituitary-adrenal axis suppression and thus no need to taper the dose prior to discontinuation.

8.2. LASERS

Lasers have become indispensible for the treatment of several sequelae, including

- Ablation of superficial proliferating and involuting vascular tissue
- Ablation of deep vascular tissue
- Treatment of atrophic scarring seen after involution
- Treatment of hypertrophic scarring seen after ulceration

Several lasers may be used, but the most common are the flashlamp pumped-dye laser (also called pulsed-dye laser, tunable dye laser) and the CO_2 laser. Before discussing these applications, an understanding of some of the basic principles is essential.

Light is a electromagnetic wave and, like all other waves, the wavelength is an extremely important measure. It is a simple measure of distance and denotes the distance from identical points on two adjacent waves or cycles (Fig. 8.2). We measure this distance in nanometers ($1nm = 1 \times 10^{-9}m$). Another important measure in light tissue interaction is time. Events take place in extremely short time intervals, and we measure these in nanoseconds (10^{-9} second), microseconds (10^{-6} second), or milliseconds (10^{-3} second).

Lasers are merely sources of extremely bright light that has certain unique properties. It is collimated (all the light travels in a straight line), coherent (all the waves are in phase with each other), and mono-chromatic (the light produced is all one or another wavelength). Each laser utilizes a specific laser medium, which in turn emits a specific wavelength of light peculiar to that laser medium. For example, CO_2

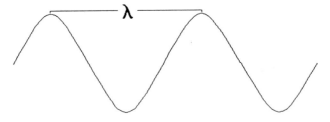

Figure 8.2
A schematic depiction of an electromagnetic wave. λ represents the wavelength.

lasers always emit light at 10,600 nm (far infrared light), and neodim-ium-yttrium-aluminium-garnet (Nd:YAG) lasers always emit light at 1,064 nm (near infrared light). Flashlamp pumped-dye lasers, on the other hand, use a dye as their active medium, and these lasers are tunable (i.e., they may be tuned over a range of wavelengths). In the context of vascular lesions, these lasers are tuned to emit light at between 585 and 600 nm (yellow light).

Light that impacts on biological tissue may be reflected, transmitted, scattered, or absorbed. These phenomena are wavelength dependent. For light to have any effect, it must be absorbed, and any of the chemical substances that make up biological tissue are capable of absorbing light. These substances are known as *chromophores*. Each chromophore has an absorbtion spectrum (i.e., a spectrum of wavelengths it will absorb to a greater or lesser degree). Because most of the hemangiomas are submucosal or subcutaneous, the most important chromophores are water, melanin, and hemaglobin, and their respective absorption spectra are shown in Figure 8.3.

The effect a particular laser will have on tissue will depend on two main factors, the wavelength of the light emitted by the laser and the length of time the tissue is exposed to the light. The wavelength is dependent on the laser medium and is by and large fixed, but the exposure time can be varied in one of several ways. Most lasers will emit a continuous beam of light when activated. This is known as *continuous wave light* (Fig. 8.4). Continuous wave light can be interrupted at regular intervals or chopped by means of an electromechanical shutter. The target will then be exposed to a series of short exposures at regular intervals determined by the shutter speed. An "on" time and an "off" time can thus be set as desired (Fig. 8.5). Unfortunately, there is a limit to the speed of the shutter, and the shortest exposure time or "on" time is around 1 msec. Modifying the power supply of the laser in such a way as to produce short bursts or pulses of a large amount of energy will result in the emission of short pulses of light. This is known as *pulsed light* (Fig. 8.6). The typical width of these pulses is very short and is in the order of microseconds (as opposed to milliseconds with chopped light). The peak power is, as a consequence, much higher (kilowatts or megawatts). The laser can also be modified optically, for example, by means of a polarizer, to once again produce pulsed light. In this instance, the pulses are even shorter (nanoseconds or picoseconds), and the peak power is much higher (gigawatts).

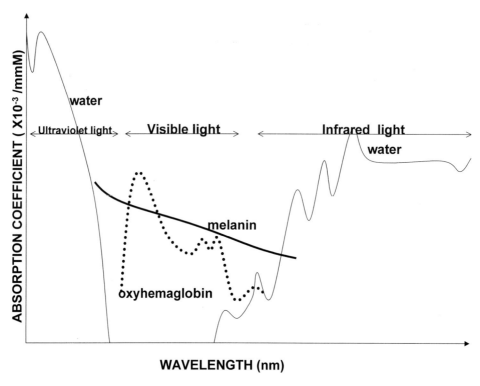

Figure 8.3
The absorption spectra of some of the important chromaphores in human skin.

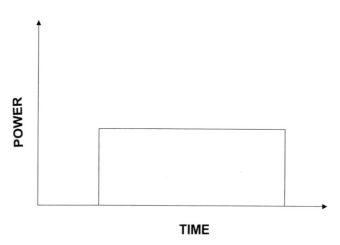

Figure 8.4
Continuous wave light. There is a continuous emission of light with little or no variation in power.

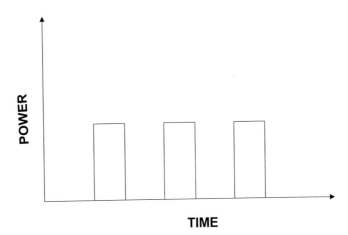

Figure 8.5
Chopped or gated continuous wave light.

The sequence of events that takes place when light is absorbed is as follows:

- Optical penetration: the light will penetrate the target and saturate it with light. This is followed by

- Thermal heating: having been saturated with light, the target will heat up and become saturated with heat. This in turn is followed by

- Thermal transmission: once the target has been thermally saturated, thermal energy will diffuse out to surrounding tissue

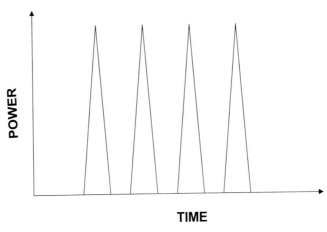

Figure 8.6
Pulsed light. Light is emitted in short pulses, each lasting several microseconds.

By limiting the exposure time to allow optical penetration and thermal heating only, one can in theory selectively damage the target chromophore. This is known as *selective photothermolysis*. In the context of hemangiomas, by using a wavelength that is selectively absorbed by hemaglobin it should be possible to destroy the vascular tissue only. The overlying skin thus should remain intact. A superficial hemangioma is usually made up of ectatic blood vessels in the papillary dermis or the mucosa. Because oxyhemaglobin is in abundance, it is used as the target chromophore. Oxyhemaglobin (HbO_2) absorbs mainly visible light, and, because the longer the wavelengths of the light the deeper the penetration within the tissue, yellow light at between 585 and 600 nm is used, as these are the longest wavelengths of light that are still strongly absorbed by HbO_2 (Fig. 10.1). The exposure time will depend on the vessel size, Fig. 10.2 and, because hemangiomas are made up of smaller vessels (100 μm or less), an exposure time of 500 msec is appropriate. The pulse width of a flashlamp pumped-dye laser falls within this requisite, and this device has become the "workhorse" laser for this application.

8.2.1. Laser Photocoagulation of Hemangiomas

Nearly every type of medical laser has been used to treat hemangiomas (Apfelberg et al., 1981; Achauer and Vander Kam, 1989; Sherwood and Tan, 1990). Unfortunately, earlier attempts led to nonselective tissue destruction. Lasers were used indiscriminately with little or no attention to the stage of the lesion or its depth. While some earlier attempts were encouraging, the results appeared to be operator dependent and were difficult to repeat (Achauer and Vander Kam, 1989; Apfelberg et al., 1981). For these reasons and the fact that hemangiomas involute, lasers were not widely accepted as a treatment modality.

The advent of the flashlamp pumped-dye laser heralded a new era in the management of vascular lesions. The correct application of this laser meant that for the first time it was possible to photocoagulate ectatic blood vessels through an intact skin or mucosa. While most of the earlier work with this laser concerned the treatment of port-wine stains, more recent work dealt with the application of this laser to the treatment of hemangiomas (Sherwood and Tan, 1990; Ashinoff and Geronemus, 1991).

Glassberg et al. (1989) and a year later Sherwood and Tan (1990) reported complete resolution of a hemangioma after several treatments with a flashlamp pumped-dye laser. Ashinoff and Geronemus reported mixed but encouraging results in a series of 10 patients. Thick prolifer-ating hemangiomas did not respond as well as involuting lesions. Unfortunately, they treated compound as well as superficial lesions in the same way. Given the limited depth of penetration of the wave-length of light they used (yellow light, 585 nm), their results were en-tirely predictable. The deeper portion of a lesion greater than 2 mm in thickness is beyond the depth of penetration of light from a flashlamp pumped-dye laser and will thus not respond. Waner et al. (1994) treated a series of children with only superficial proliferating heman-giomas and were able to effect complete resolution in all patients long before spontaneous involution would have been expected. By the same token, superficial involuting lesions will also respond. The common de-nominator is the depth of the lesion. Because light from a flashlamp pumped-dye laser (585 nm) will only penetrate to a depth of 1–2 mm, it stands to reason that only superficial lesions will respond completely. Furthermore, whatever is done to the surface of the lesion appears to have no effect on the deeper component.

Because flashlamp pumped-dye lasers are the most user friendly lasers and have been shown to give consistent reproducible results, we believe they should be regarded as the laser of choice. They can be used effectively to treat superficial hemangiomas whether they are prolifer-ating or involuting. Ulcerated hemangiomas also respond to treatment (Scheepers and Quaba, 1995). One or two treatments will usually result in healing. This mechanism is probably related to the selective effect the laser has on vascular tissue. Because the reason the ulcer persists so long is related to the rate of proliferation of the hemangioma being faster than that of the overlying skin, anything that will slow or tem-porarily stop the hemangioma from proliferating will allow the skin to "catch up" and close the wound. This only appears to be effective with superficial lesions. Ulceration involving a thicker lesion may in fact worsen. The reason for this remains unclear. Some additional factors must be operative.

The following treatment regimen is recommended:

- A standard flashlamp pumped-dye laser with a wavelength of 585 nm and a pulse width of 450–500 msec should be used. Other pa-rameters may also be effective, but at this time, in the overwhelm-ing majority of cases the above parameters were used and found to be safe and effective.

- Using a 5-mm spot size (or larger if available), the entire surface of the lesion should be treated to an end-point of purpura. This will usually require a degree of overlapping of the spots (10%–15%), and occasionally each spot will need to be treated twice during the same session to achieve sufficient purpura.

- A fluence of between 6.5 and 7.5 J/cm^2 is appropriate. When using a 7-mm spot size, we recommend decreasing the fluence by $1 \, J/cm^2$. The lower fluence is chosen for the neck and any anatomical site other than the head and face.

- If a bluish gray purpura is not seen, a higher fluence should be selected and the lesion retreated (Fig. 8.7).

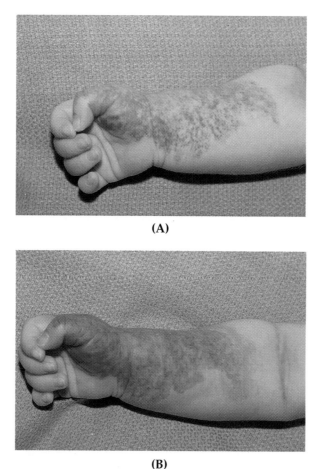

(A)

(B)

Figure 8.7
(A, B) A superficial hemangioma before and immediately after treatment with a pulsed dye laser. Note the end-point blue-gray discoloration.

• This degree of aggressive treatment will invariably result in some degree of blistering. A thin film of polysporin or some other suitable antibiotic ointment should be applied to the treated area to provide a moist environment for healing as well as to keep the area clean.

The rate of complications as a direct consequence of the treatment of hemangiomas has not been published, but, judging from the low rate of complications seen after the treatment of port-wine stains (venular malformations), this is very likely to be low (Orten et al., 1996). However, unlike vascular malformations, hemangiomas involute, and in so doing superficial hemangiomas often leave a residuum of hypopigmentation with epidermal atrophy. In addition to this, any hemangioma that has ulcerated will invariably heal with a hypertrophic scar. These confounding issues must be considered when studying the risk of complications. The rate of complications seen after treatment of port-wine stains is in the range of 1%–5% and includes hypopigmentation (especially after multiple treatments), postinflammatory hyperpigmentation (a temporary phenomenon, especially seen with types III and IV skins), atrophic scarring, and, rarely, hypertrohic scarring.

Recently, flashlamp pumped-dye lasers were modified to emit light at longer wavelengths (i.e., 590, 595, or 600 nm) and with longer pulse widths (i.e., 1,000 and 1,500 μsec). Although no data have been published using the longer wavelengths and/or the longer pulsewidths, we can safely assume that a longer wavelength will penetrate deeper and will become the preferred mode. With regard to the longer exposure times, these may be more appropriate for the treatment of the more ectatic vessels seen during involution. There may also, however, be higher risks. The longer wavelength will result in deeper penetration, but, along with this is an increased risk of hypopigmentation. The longer exposure time will address the larger vessels, but at the same time the risk of thermal damage to surrounding tissue may be higher. The reader therefore is advised to research the literature before using these more advanced lasers.

8.2.2. Skin Resurfacing

Skin resurfacing with a CO_2 laser, a technique recently popularized to treat facial rhytids, has been shown to be of benefit for treating the atrophic changes seen after involution of superficial hemangiomas of the face (Waner et al., 1996). Since the CO_2 laser emits light in the infrared re-

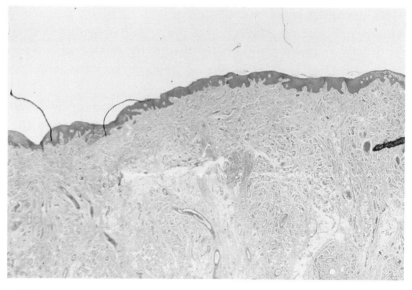

Figure 8.8
This section was taken from a hemangioma that had been resected. The area had been "resurfaced" with a CO_2 laser some 12 months prior. Note the thickened papillary dermis.

gion, it strongly is absorbed by water. This laser therefore evaporizes tissue by boiling its water. Skin resurfacing is carried out in two stages, both of which are performed at the same treatment session. The first stage involves the removal of the epidermis. A single pass with the laser will vaporize the water off the epidermis, leaving behind dessicated debris. This should be wiped away with gauze soaked in saline before any subsequent pass. A second and any subsequent pass with the laser will vaporize the water in papillary dermis, and because the water content of collagen is low and its vaporization temperature high, the heat generated will heat the collagen, which will shrink by as much as 20%. Over the course of the next 6 months, this shrunken superficial layer of collagen will eventually be extruded but appears to act as a template for a new collagen matrix that will be deposited over the treated area. In patients with rhytidosis the new collagen matrix is organized and has a thicker Grenz zone (elastin layer). In children with atrophic scarring, the net effect appears to be thickening of the papillary dermis and elimination or at least improvement of the scars. (Fig. 8.8).

A crucial element of the laser treatment is the exposure time (i.e., the length of time the tissue is exposed to the heating effect of the laser). A consideration of a time profile of the events during tissue vaporization will clarify this issue:

1. The first event after tissue exposure to light is optical penetration, (i.e., penetration of the tissue by the light). This is entirely wavelength dependent phenomenon. The wavelength of light emitted by the CO_2 laser is 10,600 nm. Light at this wavelength is strongly absorbed by H_2O, and, because the water content of skin is 96% the depth of penetration of this wavelength of light is around 60 μm.

2. The next event is vaporization. The area of tissue bathed in light is then raised to 100°C, which results in vaporization of the water from the tissue.

3. The remaining debris is dessicated, and if this tissue continues to heat it will char at 300°C. From this, thermal transmission to surrounding tissue will result.

The "thermal relaxation time" is equivalent to the time it takes for optical penetration and vaporization. If the tissue is exposed to light for an amount of time equal to or less than the thermal relaxation time, the tissue will vaporize without char formation, and hence no damage to surrounding tissue will result (char-free ablation). The exposure time is thus crucial. Whether the tissue is exposed to a pulse of light or to a rapidly moving scanned beam of light (flash-scanned light), the effect will depend on the length of time the tissue is exposed to light and not the mode of the light. The tissue effect of pulsed light and flash-scanned light is therefore equivalent provided the exposure time is the same.

Although only a small number of children have been treated, extrapolation of data from adults appears to be encouraging. Almost all patients experience improvement. The rate of complications is low, and the risk of scarring is less than 1%. The risk benefit ratio is thus high. Complications include

- Prolonged erythema
- Sepsis
- Postinflammatory hyperpigmentation
- Hypopigmentation
- Scarring

The likelihood of scarring is much higher when treating the neck because the concentration of pilosebaceous units is much lower. For this reason, we strongly advise against treating the neck with a CO_2 laser. Preliminary experience with the Er:YAG laser (erbium-yttrium-aluminum-garnet) in treating unprivileged sites such as the neck and chest is encouraging. Because light emitted by the Er:YAG laser is at least 10× more strongly absorbed by water, the depth of thermal injury beyond the zone of vaporization is much less (<5 μm). The likelihood of complications appears to be much lower, but the benefit from collagen shrinkage also appears to be less because it is currently believed that the thermal damage to collagen beyond the zone of vaporization is responsible for this shrinkage. We currently believe that the Er:YAG laser will be useful in resurfacing the neck and other unprivileged sites but that the CO_2 laser will remain the standard laser for facial resurfacing.

The regimen we suggest can be applied with any of the CO_2 lasers currently being used for cosmetic skin resurfacing. These should deliver either pulsed or continouous wave flash-scanned light (see Table 8.1). It is essential that prior to attempting resurfacing, the physician should attend a course on resurfacing and observe several cases in a preceptorship-type setting. Although the risk of complications is low, poor technique will multiply the risk severalfold.

- The first pass should remove the epidermis. Standard or default settings on most resurfacing lasers will accomplish this. (Unfortunately, many of the new generation of lasers will have a "light" setting for a more superficial peel. This will not vaporize to the dermoepidermal junction. In these cases, the "deeper" setting should be selected.)
- The debris left after this pass should be completely removed prior to

TABLE **8.1**
Appropriate Parameters for Skin Resurfacing with a CO_2 Laser

Pulsed Light		**Flashscanned Continuous Wave Light**	
Pulse width	<1 msec (1,000 μsec)	Dwell time	<1 msec (1,000 μsec)
Pulse energy	>150 mJ	Energy density	5 J/cm^2

the next pass because any remaining debris will carbonize on the next pass, and this will result in a deeper than anticipated burn.

- A second pass should then be delivered; this will heat-shrink the upper papillary layer. The effect of this will be a smoothing of scarred area.

- Once again the proteinaceous debris should be wiped away and the area should then be dressed with an occlusive hydrogel or hydrocolloid dressing.

- The occlusive dressing should be applied for at least the first 3–4 days. Thereafter the wound should be bathed three times daily to remove the exudate and petrolatum applied several times daily to keep it moist.

Re-epithelialization can be expected within 14 days and a new collagen matrix formed by 6 months. An interim period of erythema for the first 6–12 weeks and possible postinflammatory hyperpigmentation for the first 8 weeks can be expected. The full benefit of treatment will therefore not be apparent until at least 6 months post-treatment (Fig. 8.8). Occasionally a repeat treatment will be considered. This is allowable but should be avoided until after 6 months has elapsed since the first treatment.

8.2.3. Surgical Excision

Surgical excision of hemangiomas was, and to a great extent still is, considered hazardous. The risk of intraoperative hemorrhage, the difficulty in dissecting these lesions off important structures, and the fact that so many of the lesions that require intervention are in the head and neck are all confounding features. In addition to this, many of the lesions that finally come to be considered for surgery have grown considerably large. While all these considerations are valid, recent developments in technology have enabled us to minimize the risk of intraoperative hemorrhage.

8.2.3.1. Contact Laser Surgery
Traditional lasers are used in the noncontact or free beam mode. This means that there is no contact between the light transporting apparatus and tissue. The laser beam exits this apparatus (quartz fiber, wave-

guide, or articulated arm) and then impacts with the tissue. Contact laser surgery, on the other hand, implies direct contact between the apparatus transporting the light (the quartz fiber) and the tissue. In its simplest form, an optical quartz fiber can be drawn to a point and then used as a scalpel. Alternatively, a sapphire tip coupled to the fiber may be used in the same way.

The mechanism whereby light is transmitted down a quartz fiber is based on the principle of total internal reflection. All of the light entering a quartz fiber is reflected internally off the walls of the fiber. As the light enters the narrowed portion of the fiber just proximal to the tip, the angle of incident light increases until it exceeds the critical angle for total internal reflection. Once this happens, the light escapes, usually from the tip of the fiber or of the sapphire. When the tip is in contact with tissue, the light will be transmitted from the tip to the tissue with no backscatter. This will limit its depth of penetration. The light concentrating at the tip will also heat the tip well above the tissue vaporization threshold. The cutting action is therefore based on the extremely high power density of the light that vaporizes the tissue (but, due to an absence of backscatter, will not penetrate as deeply into the tissue) and the heat at the tip that will also vaporize the tissue. Transmission of both the heat and the light energy into surrounding tissue will coagulate the tissue up to 0.5 mm from the point of vaporization. This will seal small vessels and thus aid with hemostasis. The Nd:YAG laser is more efficient than a CO_2 laser, which will only seal very small vessels (0.5 mm or less), and, although there are no reliable data on the size of vessel Nd:YAG laser surgery is able to seal, it is believed to be in the region of 1 mm.

The major advantage of contact laser surgery therefore lies in its ability to seal small vessels as well as the efficiency with which it cuts. The latter may, however, be a double-edged sword. Because this device cuts so well, it will cut through tissue planes with little or no resistance. One therefore loses the tactile element of dissection and has to rely solely on tissue plane recognition during surgery.

8.2.3.2. Thermoscalpels

Merely heating the cutting edge of a scalpel blade with electrical energy will also aid with hemostasis. The cutting is achieved by both heat vaporization and the cutting effect of the scalpel blade. One is able to adjust the temperature of the blade from room temperature up to 320°C and thereby the speed of vaporization as well as the degree of coagulation. The higher the temperature, the greater the degree of coagulation and the more efficient the hemostasis. Vaporization will also be more rapid with a higher temperature.

The extent of the thermal changes in the tissue surrounding the incision will depend on the temperature of the scalpel as well as the speed with which the incision was made. The higher the temperature of the scalpel and the slower the cut, the greater the degree of transmission of the thermal energy to the surrounding tissue. This in turn means greater thermal damage to surrounding tissue and also greater coagulation and therefore better hemostasis. When raising a thin flap, one would obviously choose a lower temperature to prevent flap necrosis, whereas a higher temperature would be safe for thicker flaps. In addition to sealing vessels during incising, the scalpel can also be used as a coagulator. By placing the flat edge of the blade against a bleeding surface the surface will heat, and this in turn will seal the smaller vessels.

During surgical dissection, some degree of resistance is encountered between the various tissue planes. One therefore relies on both tactile and visual recognition of the planes. Needless to say, recognition of the tissue planes is essential to good technique. The use of thermoscalpels is thus extremely useful because it assists with hemostasis without diminishing the degree of resistance between the planes even though bleeding may have diminished visual recognition.

8.2.3.3. Surgical Techniques

Two of the most important aspects of surgery are the control of blood loss and the location of the correct plane for dissection of the lesion. In many ways, these are interdependent. Good control of bleeding is essential to recognition of the correct surgical plane, and by remaining within this plane one is able to minimize blood loss. Hemostasis is therefore the key to a successful resection. The surgeon should make use of a thermoscalpel and/or contact laser surgery, as well as electrocautery. Bipolar electrocautery is preferred when dissecting around the facial nerve and monopolar electrocautery when no vital/delicate structures are in the vicinity of the lesion. Because tissue plane recognition is so important, the maintenance of a clean dry field is helpful. The surgeon should therefore deal with any bleeding that occurs when it occurs and not at some later stage.

During proliferation, a hemangioma may expand rapidly and con-
dense a soft tissue plane around it. This plain is relatively avascular
and, once accessed, is easy to follow. Because hemangiomas will de-
form but rarely cross tissue planes, staying within this plane will sim-
plify the task of removing the lesion (Fig. 8.9A, B, C, D). Unfortunately,
as involution progresses, this plane becomes less and less distinct until
in late involution it is hardly discernible. The lesion will by now have
transformed to fibro-fatty tissue, indistinguishable from surrounding
subcutaneous fat. Therefore, contrary to widely held belief, resecting a
lesion earlier rather than later has its advantages. A corollary to this is
the fact that because the hemangioma is involuting it is not essential to
remove the entire lesion, especially in cases where removal of the entire
lesion is likely to result in significant morbidity. Indeed, in some cases
it is even preferable to leave some lesion behind because it will trans-
form to much needed fibro-fatty tissue. Surgeons should use their judgment when making these decisions.

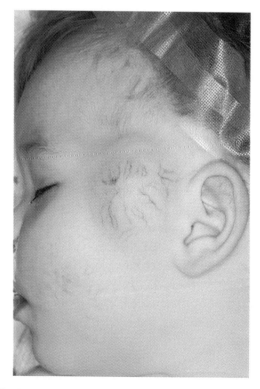

Figure 8.9A, B, C, D
Surgical removal of a hemangioma with the aid of a thermoscalpel. Note the presence of a false capsule and the relatively avascular dissection plane. The entire lesion was superficial to the parotid fascia.

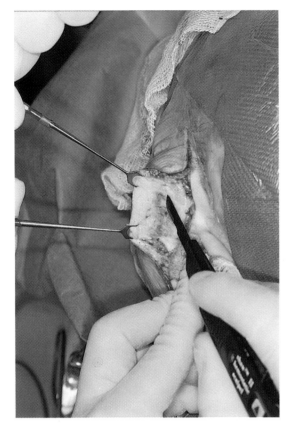

Figure 8.9B

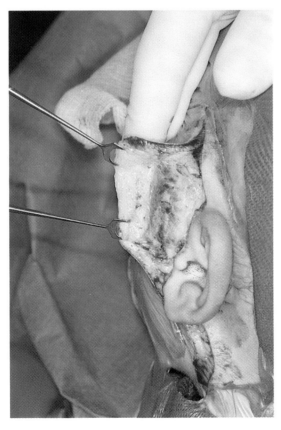

Figure 8.9C

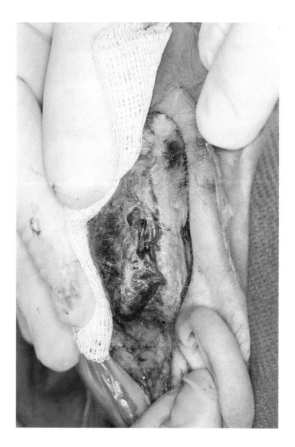

Figure 8.9D

Another advantage of early removal that should be considered is what has become known as the *skin expander effect*. During the early stages of the life cycle of a hemangioma, a large lesion will distend, surrounding tissues, including skin. If the lesion is compound and there is considerable skin involvement, the expanded uninvolved excess skin surrounding this can often be used to achieve primary closure. This effect may be lost during late involution because, as the lesion shrinks, a point is reached where the amount of expanded skin is no longer sufficient to achieve primary closure. This effect, whenever present, therefore must be taken into account when determining the optimum time for intervention.

REFERENCES

Apfelberg, D.B., Greene, R.A., Maser, R.: Results of argon laser exposure of capillary hemangiomas. Plast. Reconstr. Surg. 67:188–193, 1981.

Achauer, B.M., Vander Kam, V.M.: Capillary hemangioma (strawberry mark) of infancy: Comparison of argon and Nd/YAG laser treatment. Plast. Reconstr. Surg. 84:60–69, 1989.

Ashinoff, R., Geronemus, R.G.: Capillary hemangiomas and treatment with the flashlamp pumped-dye laser. Arch. Dermatol. 127:202–205, 1991.

Azzolini, A., and Nouvenne, R.: Nuove prospettive nella terapia degli angiomi immaturi dell'infanzia. 115 lesioni trattate con infiltrazioni intralesionali di triamcinnolone acetonide. Ateneo Parmense (suppl.) 41:51, 1970.

Bartoshesky, L., Bull, M., and Feingold, M.: Corticosteroid treatment of cutaneous hemangiomas: How effective? Clin. Pediatr. 17:625–638, 1978.

Brouty-Boye D., and Zetter, B.R.: Inhibition of cell motility by interferon. Science 208:516–518, 1980.

Brown, S.H., Neerhaut, R.C., Fonkalsrud, E.W.: Prednisone therapy in the management of large hemangiomas in infants & children. Surgery 71:168, 1972.

Brown, B.Z. and Huffaker, G.: Local injection of steroids for juvenile hemangiomas which disturb the visual axis. Ophthalmic Surg. 13:630, 1982.

Cohen, S., and Wang, C.: Steroid treatment of hemangioma of the head and neck in children. Ann. Otol. 81:584–590, 1972.

Crum, R., Szabo, S., and Folkman, J.: A new class of steroids inhibits angiogenesis in the presence of heparin or a heparin fragment. Science 230:1375–1376, 1985.

Deans, R., Harris, G., and Kivlin, J.: Surgical dissection of capillary hemangioma: An alternative to intralesional corticosteroids. Arch. Ophthalmol. 110:1743–1747, 1992.

Droste, P., Ellis, F., Sondhi, N., and Helveston, E.: Linear subcutaneous fat atrophy after cortiscosteroid injection for periocular hemangioma. Am. J. Ophthalmol. 105:65–69, 1988.

Edgerton, M.: The treatment of hemangioma: With special reference to the role of steroid therapy. Ann. Surg. 183:517–532, 1976.

Elsas, F., and Lewis, A.: Topical treatment of periocular capillary hemangioma. J. Pediatr. Ophthalmol. Strabis. 31:153–156, 1994.

Enjolras, O., Riche, M., Merland, J., and Escande, J.: Management of alarming hemangiomas in infancy: A review of 25 cases. Pediatrics 85:491–498, 1990.

Ezekowitz, R., Phil, D., Mulliken, J., and Folkman, J.: Interferon ALFA-2a therapy for life-threatening hemangioma of infancy. N. Engl. J. Med. 326:1456, 1992.

Feldman, D., Goldstein, A.L., Cox, D.C., and Grimley, P.M.: Cultured human endothelial cells treated with recombinant leukocyte A interferon: Tubuloreticular inclusion formation, antiproliferative effect, and 2', 5'-oligoadenylate synthetase induction. Lab. Invest. 58:584–589, 1988.

Folkman, J.: Successful treatment of an angiogenic disease. N. Engl. J. Med. 320:1211–1212, 1989.

Fost, N., and Esterly, N.: Successful treatment of juvenile hemangioma with prednisone. J. Pediatr. Am. Acad. Dermatol. 72:351–357, 1968.

Friesel, R., Komoriya, A., and Maciag, T.: Inhibition of endothelial cell proliferation by gamma-interferon. J. Cell. Biol. 104:689–696, 1987.

Groopman, J.E., Gottlieb, M.S., and Goodman, J.: Recombinant alpha-2-interferon therapy for Kaposi's sarcoma associated with acquired immunodeficiency syndrome. Ann. Intern. Med. 100:671, 1984.

Gunn, T., Reece, E., Metrakos, K., et al.: Depressed T cell following neonatal steroid treatment. Pediatrics 6:61–67, 1981.

Guyer, D., Tiedeman, J., Yannuzzi, L., Slakter, J., Parke, D., Kelley, J., Tang, R., Marmor, M., Abrams, G., Miller, J., and Gragoudas, E.: Interferon-associated retinopathy. Arch. Ophthalmol. 111:1, 1993.

Heyns, A., Eldor, A., Vlodavsky, I., Kaiser, N., Fridman, R., and Panet, A.: The antiproliferative effect on interferon and the mitogenic activity of growth factors are independent cell cycle events: Studies with vascular smooth muscle cells and endothelial cells. Exp. Cell Res. 161:297–306, 1985.

Hiles, D., and Pilchard, W.: Corticosteroid control of neonatal hemangiomas of the orbita and ocular adnexa. Am. J. Ophthalmol. 71:1003–1008, 1971.

Katz, H.: Thrombocytopenia associated with hemangioma: Critical analysis of steroid therapy. Proceedings of the XI International Congress of Pediatrics, Tokyo, p. 336, 1965.

Katz, H., and Askin, J.: Multiple hemangioma with thrombocytopenia. An unusual case with comments on steroid therapy. Arch. Pediatr. Adolesc. Med. 115:351–357, 1968.

Kushner, B.: Local steroid therapy in adnexal hemangioma. Ann. Ophthamol. 11:1005–1009, 1979.

Kushner, B.: Treatment of periorbital infantile hemangioma with intralesional corticosteroid. Plast. Reconstr. Surg. 76:517–526, 1985.

Loughnan, M., Elder, J., and Kemp, A.: Treatment of a massive orbital-capillary hemangioma with interferon alfa-2b: Short term results. Arch. Ophthalmol. 110:1, 1992.

Mulliken, J., Boon, L., Takahashi, K., Folksman, J., and Esekowitz, A.: Pharmacologic therapy for endangering hemangiomas. Curr. Opin.: Dermatol. 109–113, 1995.

Ohlms, L., McGill, T., Jones, D., and Healy, G.: Interferon alfa-2A therapy for airway hemangioma. Ann. Otol. Rhinol. Laryngol. 103:1, 1994.

Oleszak, E., and Inglot, A.: Platelet-derived growth factor (PDGF) inhibits antiviral and anticellular action of interferon in synchronized mouse or human cells. J. Interferon Res. 37–48, 1980. Vol 1

Orten, S.S., Waner, M., Flock, S., Roberson, P.K., Kincannon, J.: Port-wine stains. An assessment of 5 years of treatment. Arch. Otolarynyol. Head and Neck Surg. 122: 1174–1179, 1996.

Real, F.X., Ottegen, H.F. and Krown, S.E.: Kaposi's sarcoma and the acquired immunodeficiency syndrome: treatment with high & low doses of recombinant leukocyte A interferon. J. Clin. Oncol. 4:544, 1986.

Rios, A., Mansell, P.W., Newell, G.R., Reuben J. M., Hersh, E.M. and Gutterman, J.U.: Treatment of acquired immunodeficiency syndrome related kaposi's sarcoma with lymphoblastoid interferon. J. Clin. Oncol. 3:506, 1985.

Ruttum, M., Abrams, G., Harris, G., and Ellis, M.: Bilateral retinal embolization associated with intralesional corticosteroid injection for capillary hemangioma for infancy. J. Pediatr. Ophthalmol. Strabis. 30:4–7, 1993.

Sadan, N., and Wolach, B.: Treatment of hemangiomas of infants with high doses of prednisone. J. Pediatr. 128:141–146, 1996.

Sasaki, G., Pang, C., and Wittliff, J.: Pathogenesis and treatment of infant skin strawberry hemangiomas: Clinical and in vitro studies of hormonal effects. Plast. Reconstr. Surg. 73:359–370, 1984.

Scheepers, J.H. and Quaba, A.A.: Does the pulsed tunable dye laser have a role in the management of infantile hemangiomas? Observations based on three years' experience. Plast. Reconstr. Surg. 95:305, 1995.

Sherwood, K.A. and Tan, O.T.: Treatment of a capillary hemangioma with a flashlamp pumped dye laser. J. Am Acad. of Dermatol. 22:136, 1990.

Shorr, N., and Seiff, S.: Central retinal artery occlusion associated with periocular corticosteroid injection for juvenile hemangioma. Ophthalmic Surg. 17:229–231, 1986.

Sloan, G., Reinisch, J., Nichter, L., et al Intralesional corticosteroid therapy for infantile hemangioma. Plast. Reconstr. Surg. 83:459–466, 1989.

Soumekh, B., Adams, G., and Shapiro, R.: Treatment of head and neck hemangioma with recombinant interferon alpha 2B. Ann. Otol. Laryngol. 105:1996.

Sutula, F., and Glover, A.: Eyelid necrosis following intralesional corticosteroid injection for capillary hemangioma. Ophthalmic Surg. 18:103–105, 1987.

Tsuroka, N., Sugiyama, M., Tawaragi, Y., et al. Inhibition of in vitro angiogenesis by lymphotozin and interferon-gamma. Biochem. Biophis. Res. Commun. 155:429–435, 1988.

Vesikari, T., Nuutila, A., and Cantell, K.: Neurologic sequelae following interferon therapy of juvenile laryngeal papilloma. Acta. Paediatr. Scand. 77:619–622, 1988.

Waner, M., Suen, J.Y., Dinehart, S. and Mallory, S.B.: Laser photocoagulation of superficial proliferating hemangiomas. J. Dermatol. Surg. Oncol. 20:43, 1994.

Waner, M.: Laser resurfacing and the treatment of involuting hemangiomas. Laser Surg Med. Suppl. 8. Abstract 219, 40, 1996.

Weiss, A.: Adrenal suppression after corticosteroid injection of periocular hemangioma. Am. J. Ophthalmol. 107:518–522, 1989.

Wilfingseder, P., and Propst, A.: Klinische und morphologische Aspekta zur Cortison-Therapie von Hamangiomen. Chir. Plastica (Berl.) 1:109, 1972.

Wyman, L., Fulton, G., and Shulman, M.: Direct observations on the circulation in the hamster cheek pouch in adrenal insufficiency and experimental hypercorticalism. Ann. N.Y. Acad. Sci. 56:643–658, 1953.

Zak, T.A., and Morin, J.D.: Early local steroid therapy of infantile eyelid hemangiomas (local steroid therapy of lid hemangiomas) J. Pediatr. Ophthalmol. Strabismus 18:25, 1981.

Zarem, H., and Edgerton, M.: Induced resolution of cavernous hemangioma following prednisolone therapy. Plast. Reconstr. Surg. 39:76–83, 1967.

Zweifach, B., Shorr, E., and Black, M.: The influence of the adrenal cortex on behavior of terminal vascular bed. Ann. N.Y. Acad. Sci. 56:626–633, 1953.

The Treatment of Hemangiomas

MILTON WANER, M.D., F.C.S. (S.A.) AND
JAMES Y. SUEN, M.D. F.A.C.S.

Current dogma dictates that most hemangiomas should be left untreated. However, contrary to widely held belief, a cosmetically unacceptable result is often left at the end of involution (Finn et al., 1983). If
this is so, the consequences of benign neglect are unjustified. The psychosocial trauma of a readily visible hemangioma as well as the disappointment of an unacceptable result and the additional trauma of corrective surgery between the ages of 5 and 15 years all mitigate for a
re-examination of the policy of benign neglect.

Two of the most frequently asked questions concern the degree and
the duration of involution. Will the hemangioma look completely normal,
and how long will it take? The answer to the second of these questions
has been well documented. Fifty percent will have completed involution
by age 5 years and 70% by age 12 (Bowers et al., 1960). Unfortunately, the
degree of involution had, until recently, evaded close scrutiny. The suppo-
sition that the vast majority will completely involute has, and still

Hemangiomas and Vascular Malformations of the Head and Neck, Edited by Waner, M.D. and Suen, M.D.
ISBN 0471-17597-8 © 1999 Wiley-Liss, Inc.

TABLE **9.1**
Possible Outcomes in the Life Cycle of an Hemangioma

Age at Completion of Involution	Good Outcome	Poor Outcome
<6 yr.	62%	38%
>6 yr.	20%	80%
	Ave. = 41%	Ave. = 56%

remains, widely held. Finn et al. (1983) have clarified this issue. They defined complete involution as meaning just that: "no redundant skin scar or telangiectasia." In their analysis of a large number of cases, they concluded that 80% of lesions that *had not* involuted by age 6, left a significant cosmetic deformity, whereas 38% of those had *had* involuted by age 6 left a significant cosmetic deformity (see Table 9.1). Thus, if approximately half of the lesions complete involution by age 6, then the majority (59%) will leave a significant cosmetic deformity. In light of this, the answer to the first question should be "no" in the majority of cases (Figs. 9.1–9.4).

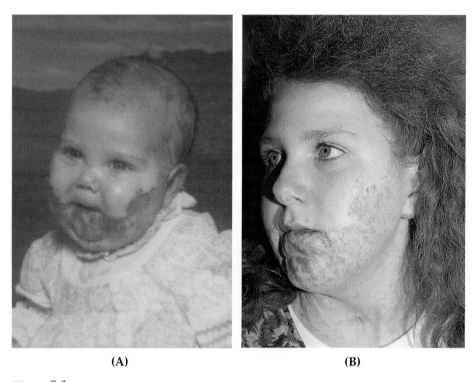

(A) (B)

Figure 9.1
(A) An infant with a superficial "beard" hemangioma (involving the V3 dermatome). **(B)** The same child at 16 years of age. Note the atrophic scarring, residual telangiectasia, and residual fibro-fatty tissue.

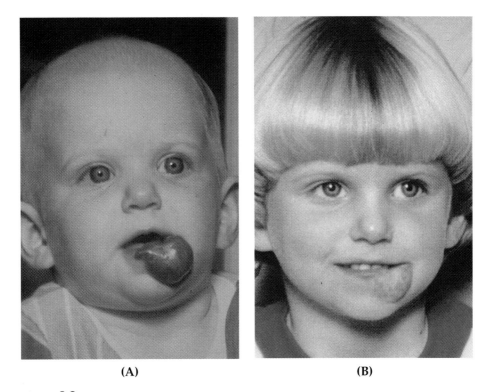

Figure 9.2
(A) An infant with a hemangioma of her lower lip. **(B)** The same child at 5 years of age. Note the residual fibro-fatty tissue and epidermal atrophy.

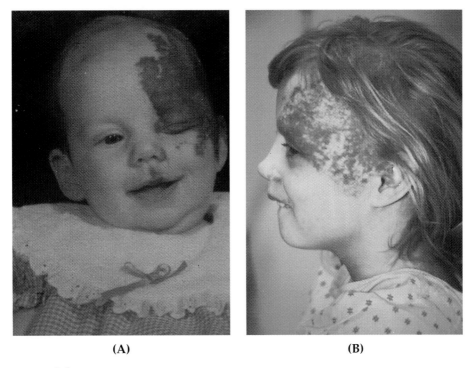

Figure 9.3
(A) An infant with a superficial hemangioma. **(B)** The same child at 5 years of age with residual telangiectasia.

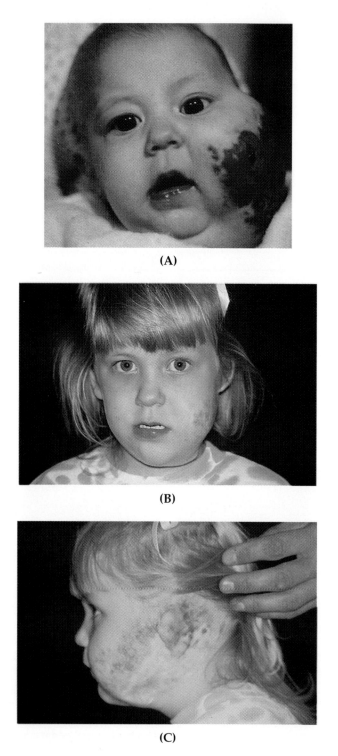

Figure 9.4
(A) An infant with a large ulcerating parotid hemangioma. **(B)** Frontal view of the same child at 10 years of age. **(C)** Lateral view of the same child at 10 years of age.

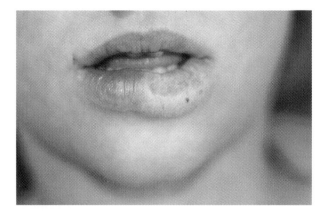

Figure 9.5
A 9-year-old child with a residuum of fibro-fatty tissue and hemangioma. This lesion has ulcerated during infancy, causing the atrophic scar.

Our standard response that the lesion will completely disappear within the first few years of life is therefore erroneous and misleading (Figs. 9.5–9.7). Superficial hemangiomas replace the papillary dermis and, as they proliferate, stretch the overlying epidermis. In addition to this, mast cell degranulation, which probably takes place during proliferation, may

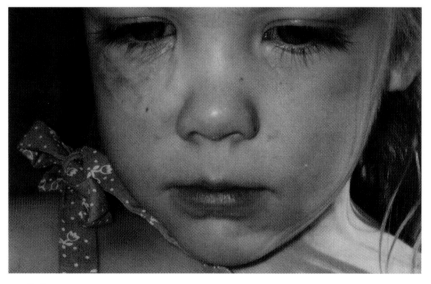

Figure 9.6
A 4-year-old child with epidermal atrophy of the lower eyelid. Despite complete and rapid involution, an atrophic scar is still evident.

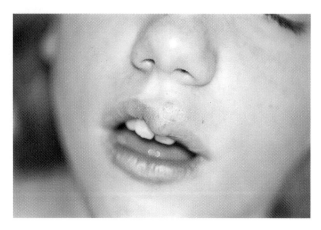

Figure 9.7
A 10-year-old child with a residuum of epidermal atrophy and fibro-fatty tissue after involution of a compound hemangioma.

well cause elastolysis. Either one or both of these factors will result in an atrophic scar. The result of involution of a superficial hemangioma is thus likely to be epidermal atrophy and telangiectasia. A subcutaneous hemangioma, on the other hand, is likely to leave a mass of residual fibro-fatty tissue, whereas compound lesions are likely to leave a residual fibro-fatty mass covered with atrophic skin and telangiectasia. (Waner et al., 1992).

Before sentencing a patient to several years of benign neglect, the psychosocial consequences of this action should be considered (Figs. 9.8–9.10). Unfortunately, this is one of the most neglected aspects of the management of hemangiomas. Several of the major textbooks fail to even mention this. Perhaps our embarrassment at having to deal with something as trivial as the "cosmetic" appearance of the child is in part responsible when, in reality, our neglect can lead to profound psychosocial trauma during the formative years of the child's personality, the consequences of which may well be irreversible. A child first develops self-awareness at between 18 and 24 months of age. From then on, each major period of development will be affected by the presence of a hemangioma. Likewise, the effect on the child's parents, as well as any siblings, can be profound. Guilt, anger, disappointment, and over protectiveness are all common.

Lastly, in re-evaluating our policy of benign neglect, we must consider whether recent developments in medicine are of sufficient benefit to warrant a change. The policy of benign neglect evolved as a conse-

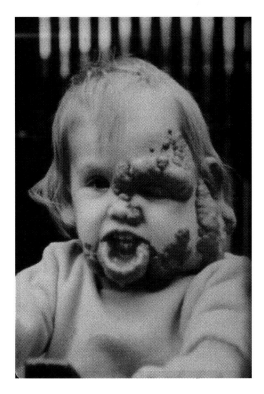

Figure **9.8**
An 18-month-old child with extensive facial hemangiomas.

Figure **9.9**
A 3-year-old child with an upper lip hemangioma. At 3 years of age this child already knew she was disfigured. Her hand would cover her lip every time she met
a stranger.

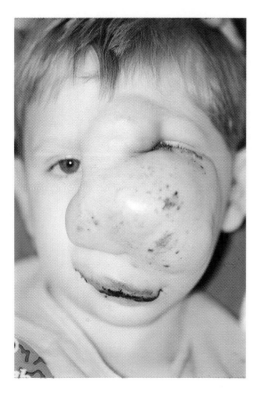

Figure 9.10
A 5-year-old boy with a massive paranasal lesion. This child was sent home from Sunday School because his disfigurement adversely affected the other children in his class.

quence of the astute clinical observations of Lister in 1938 and became firmly established over the next two to three decades (Margileth and Museles, 1965). Needless to say, during that era there were few if any other options. Since then, much has changed. We now have drugs that can shrink hemangiomas, lasers that can selectively destroy vascular tissue, and surgical instruments that almost eliminate the hazard of intraoperative hemorrhage. We thus have the means to change the natural history of hemangiomas. Therefore, in consideration of the following points, a change in our policy is clearly warranted with the provision that whatever we do should offer a distinct advantage over "benign neglect":

1. Only 40% of all hemangiomas involute completely; the remaining 60% will benefit from some form of corrective surgery whether it be laser treatment, corrective surgery, or both.

2. We are much better at predicting the final outcome of hemangiomas and can therefore offer early intervention to those who are less likely to do well.

3. The psychological trauma is considerable and probably irreversible.

4. We now have the means to safely treat hemangiomas.

9.1. STAGES IN THE LIFE CYCLE

An approach to the management of hemangiomas must, of necessity, take into account the different stages in the metamorphosis of these lesions. A proliferating hemangioma is both clinically and pathophysiologically different from an involuting hemangioma, and, by the same token, so too is a lesion in early involution quite different from one in late involution.

During proliferation the lesion is cellular (see Fig. 2.9). Tubules of plump proliferating endothelial cells surrounding narrow vascular channels are typical. Mast cells are abundant, and, despite the fact that there appears to be no organization, reticulum staining reveals a well-organized matrix. As the term implies, a *proliferating lesion* is in an aggressive, active phase of growth. Endothelial cells are actively proliferating; as a consequence, frequent mitoses are seen. Proliferation will begin to slow by 6 months and will have completely ceased by 10–14 months of life. This marks the commencement of the sometimes long, protracted course of involution. In contrast to proliferation, flatter, less active endothelial cells surrounding more obvious, ectatic vascular channels are now seen. A fibroareolar stroma becomes apparent, and, as the number of these ectatic vessels diminishes, they are replaced by more of this stroma. Late involution is the final stage in the metamorphosis. Flat, inactive endothelial cells surrounding ectatic vascular channels and an abundance of stromal fibro-fatty tissues are characteristic features (see Fig. 2.15). We have thus progressed from a predominantly cellular, actively proliferating lesion to an involuting, inactive vascular lesion.

9.2. TREATMENT MODALITIES

Enormous advances in medical science have occurred in the past decade. We now have thermoscalpels capable of reducing blood loss during surgery and lasers capable of selectively destroying vascular

tissue through intact skin or mucosa. We also have drugs capable of inhibiting angiogenesis and magnetic resonance imaging (MRI) capable
of both diagnosing the nature of a lesion and demarcating it's extent.
Because all of these modalities are useful, with some only at a particular stage in the life cycle, a logical approach to the management of hemangiomas would be to consider the management of each stage in the
life cycle of the hemangioma and the treatment modalities pertaining
to that stage separately. In so doing, we also hope to avoid the contro-
versies surrounding some of these treatments.

9.3. GENERAL PRINCIPLES

Regardless of the stage or site of the lesion, certain potential sequelae
should always be kept in mind when managing a patient with one or
more hemangiomas (Table 9.2). The infant should be seen frequently
and examined thoroughly, especially during the first 3 months. The
child's growth and development should be monitored since failure to
thrive is sometimes seen with large hemangiomas or multiple small lesions, especially visceral and or hepatic involvement. A large oropha-
ryngeal or hypopharyngeal lesion may increase the work of breathing
and this may also result in failure to thrive. This obstruction may also
lead to right ventricular hypertrophy and, in extreme cases, cor pulmonale. A large hemangioma may also precipitate high-output cardiac
failure through an increase in the circulating blood volume, which in
turn causes ventricular hypertrophy and eventually congestive cardiac
failure (Vaksman et al., 1987; Cooper and Bolande, 1965; Dupont et al.,
1977). An increased circulating blood volume may also result in hemodeviation with a disturbance of the hemodynamic equilibration.

TABLE **9.2**
Clinical Sequelae Associated with Hemangiomas

Failure to thrive
Upper airway obstruction
High-output cardiac failure
Cor pulmonale
Kasabach-Merritt syndrome

This may result in hypoperfusion of other areas, especially cerebral tissue, which could predispose to seizures. Signs and symptoms of hemodeviation include feeding difficulties, dyspnea, cyanosis, tachycardia, an ejection systolic murmer, or hepatomegaly. Because children with hepatic hemangiomas have high mortality rates (usually from cardiac failure), early detection is helpful so that these lesions may be treated aggressively (Larcher et al., 1981; Pereyra et al., 1982; Vaksman et al., 1987). Unfortunately, there are no reliable indicators of visceral involvement; therefore, once again, repeated clinical examination is essential. An exception to this is in children with multiple small cutaneous lesions. This condition is known as *disseminated* or *diffuse* neonatal hemangiomatosis. Visceral involvement should be actively sought because it is more likely in these cases. Both ultrasound and MRI are useful investigations in this regard. An altered cry, stridor, a croupy cough, or feeding difficulties are suggestive of a subglottic hemangioma. Preauricular skin involvement as well as the lower lip and chin is often associated with hypopharyngeal airway embarrassment.

9.4. TREATMENT ACCORDING TO STAGE

9.4.1. General Principles

As the hemangioma proceeds through the stages of its life cycle, the transition from stage to stage is gradual, but each has its own distinct clinical and histological features. Because each of the treatment modalities can be used in more than one stage, a logical approach to the management of hemangiomas must consider each phase separately. Although there is considerable overlap during the transition from one stage to the next, this format will be used throughout this chapter. The preferred treatment will thus depend on the stage at which the patient is first seen.

Life- or sight-threatening hemangiomas should always be dealt with as needed, regardless of the stage of the lesion. With regard to the rest, we recommend active intervention in all obvious head and neck or readily apparent lesions. Our aim is to prevent disfigurement and in so doing prevent the psychosocial trauma these children experience. The timing of intervention should always take into account the natural history of hemangiomas, as well as the particular pattern the lesion under consideration seems to be following. In general, treatment should be aggressive during the first year of life (i.e., during proliferation) and again later at around 3 years of age if this is necessary. With early aggressive management we hope to either completely eradicate the lesion or at least stunt its growth so that by the time proliferation is complete we are left with a much smaller, uncomplicated, or less disfiguring lesion. By the end of the proliferative phase, medical intervention (apart from interferon) is no longer effective. In addition, the hemangioma has begun to involute. Benign neglect is appropriate at this stage. Fairly soon it will become apparent whether or not the hemangioma is involuting rapidly. Should this not be the case, and in some cases even if this is the case, we again face the decision of whether or not we should intervene. Several factors will color this decision, the most important of which are the degree of disfigurement, the psychological trauma, and the speed with which the hemangioma is involuting. Statistically, lesions that involute rapidly are more likely to do well and can often be left alone. On the other hand, the "slow involuters" almost always require corrective surgery, and, unless contraindicated, this should begin early because several procedures are often necessary. Our aim is to complete this corrective surgery as soon as possible because of the early age during at which the child develops self-esteem. Timely intervention should therefore prevent or at least minimize the psychosocial trauma. Others factors that will affect our decision include the anatomical location of the lesion and the severity of the disfigurement. The decision to intervene should always be made with the full participation of the parents and all the members of the vascular anomalies team and must take into account the parental attitude as well as the psychosocial disposition of the child.

9.4.2. Early Proliferation

A localized area of telangiectasia or an erythematous macule are the most common signs of an early superficial proliferating hemangioma (Hidano and Nakajima, 1972). Proliferation may also commence in the deeper reticular dermis or, as is commonly seen, in both superficial and deep areas. Given the limited depth of penetration of yellow light (0.5 mm at 578 nm and 600μ at 585 nm), laser photocoagulation will only be effective in treating the most superficial of lesions (Waner and Suen,

1992; Waner et al., 1994, Garden et al., 1992). In these cases, one is able to completely destroy a proliferating hemangioma (Figs. 9.11, 9.12).

Using a 5-mm spot size and overlapping spots by about 10%–20%, the entire lesion should be treated at between 6.5 and 8.0 J/cm² (5-mm spot size) until an end point of uniform purplish gray discoloration of the entire lesion is achieved (Fig. 9.13). (This represents coagulation of the hemoglobin.) On occasion and especially with thicker lesions, it may be necessary to photocoagulate an area twice during the same treatment to achieve this end point. Treatment should commence at the earliest sign of a hemangioma and be repeated at 4–6 weekly intervals until the lesion has been completely eliminated (Garden et al., 1992). If proliferation has been entirely cutaneous, complete resolution will be possible. Up to six treatments may be required to achieve this (Waner and Suen, 1992). On the other hand, a hitherto undetected area of subcutaneous

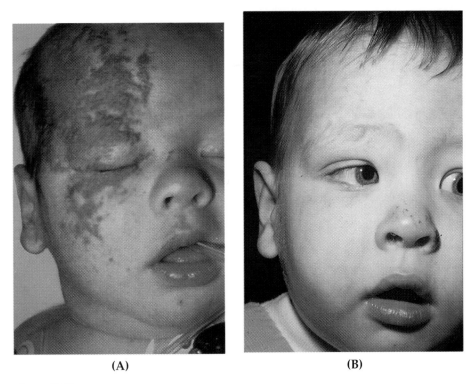

(A)   (B)

Figure 9.11
(A) An infant with a extensive rapidly proliferating superficial forehead hemangioma. (B) The same child at 1 year of age after six treatments with a flashlamp pumped-dye laser.

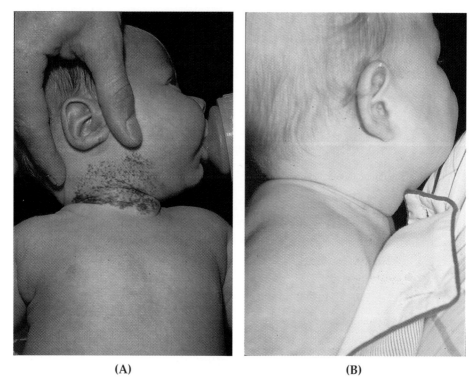

(A)   **(B)**

Figure **9.12**
(A) An infant with a rapidly proliferating superficial cervical hemangioma. **(B)** The same child after two treatments with a flashlamp pumped-dye laser.

proliferation may become apparent after the superficial component has been eliminated. This deeper component will continue to proliferate unabated. Previous claims that treatment of the surface will slow proliferation or induce a more rapid involution of the residual hemangiomas are anecdotal and remain unsubstantiated.

Papular lesions, raised greater than 1 mm above the surface of the skin, have not been found to respond well to treatment with a conventional flashlamp pumped-dye laser (Garden et al., 1992). This is entirely predictable. The depth of penetration of light at 585 nm is limited to around 1 mm, and the short 450 μsec exposure time will prevent thermal transmission to the deeper tissue. Preliminary experience in treating these with a new generation of flashlamp pumped-dye lasers that emit light with a longer pulse width (1.5 msec) and a longer wavelength (590–600 nm) has been encouraging. The longer wavelength

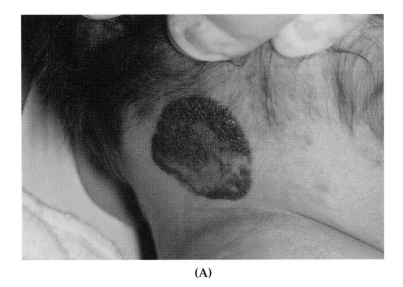

(A)

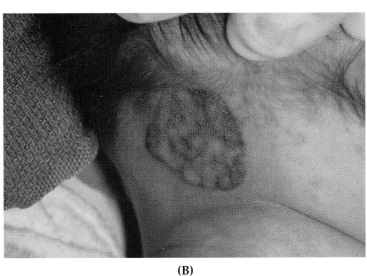

(B)

Figure 9.13
(A) A superficial occipital hemangioma before treatment. (B) The same lesion immediately after treatment. Note the bluish gray appearance. This is an appropriate end point.

(590–600 nm) and the greater thermal energy possible as a consequence of the longer exposure time with this laser will result in deeper photocoagulation, thereby eliminating these lesions more readily. Although this may seem advantageous, the risks of scarring and hypopigmentation may well be increased. Unfortunately, no data on this are available. Despite the increased depth of penetration afforded by the longer wavelength, there is a limit to the depth of the effect due entirely to the depth of penetration of this wavelength of light. This longer wavelength will at best add an additional 4–5 mm to the depth of effective coagulation.

Early proliferation of a deeper lesion will present as a vague bluish discoloration or a frank mass. Because the lesion was in most cases not present at birth and will continue to proliferate (during the first year), the diagnosis (and unfortunately the lesion) will in most cases be obvious. The object of treatment is to stunt the growth of the lesion or if at all possible, eliminate it completely. By so doing, we hope that at the end of proliferation the lesion is much smaller than it would have been and, at best, barely visible. Although at this point we still do not know whether this will have any effect on the degree of involution or its speed, a smaller lesion is less obvious, easier to manage and may not require any further treatment.

A course of steroids will invariably shrink the lesion, and the child should remain on a sufficient dose for a sufficient period of time to have any lasting effect on the lesion. This may mean as long as several months (see Chapter 8 for regimen).

9.4.3. Late Proliferation

By this stage, the hemangioma has usually acquired a thickness and/or depth beyond the depth of penetration of yellow light. This unfortunately precludes effective treatment with a yellow light laser. Steroids will on the other hand be able to shrink the lesion, and, provided they are given for a long enough period and at an appropriate dosage, the proliferative phase will be stunted (Sadan and Wolach, 1996). Our aim once again is to ensure that at the end of proliferation the lesion has either completely involuted or is much smaller and less disfiguring than it would have been without treatment. The preferred route of administration is oral. Intralesional steroids are reserved for small localized facial lesions. With regard to periocular hemangiomas, the severity of complications must be considered when treating with intralesional steroid injections. We thus prefer the oral route for these lesions as well even though the majority of ophthalmologists still use intralesional steroids.

The physician may, on the other hand, elect a more aggressive approach. Because steroids will only be effective during proliferation, as the lesion advances in maturity, we face a situation of diminishing response with regard to steroids. Therefore, when faced with a lesion that has failed to respond to steroids or, despite our best efforts, is still unacceptably disfiguring, surgery, with or without laser treatment, is an alternative. This is especially true for children with a lesion in whom surgery is inevitable. These include:

- Lesions that have ulcerated and subsequently scarred (Fig. 9.5, 2.17, and 2.18)
- Life or site-threatening hemangiomas (Figs. 9.14, 2.27, 2.28, 2.29)
- Hemangiomas in certain anatomical sites associated with a poorer prognosis (nasal tip, lip, paranasal) (Figs. 9.10, 9.15)

Benign neglect in the form of parental reassurance and careful follow up should be dispensed for an uncomplicated lesion involving any other site, with the proviso that the lesion is not cosmetically obvious. A full explanation, offered as often as is necessary, is essential to establishing a good rapport with the parents. This, together with repeated clinical examinations, will be reassuring to the parents and will enable the physician to keep a close eye on the hemangioma.

This form of benign neglect should continue until 3–4 years of age. By this stage the process of involution should be well underway, and it should by now be apparent whether or not involution will be complete before 6 years of age. Should this be the case, benign neglect should continue because, in 60% of these cases, spontaneous involution will leave an excellent cosmetic result. On the other hand, should the converse be true, corrective treatment should be considered because 80% of these cases will result in an unfavorable outcome. The parents should be made aware of these statistics. Despite the potentially poor outcome, they may occasionally still elect to continue with benign neglect because the lesion is not obvious and therefore not causing an immediate problem.

Regardless of the treatment plan, by the end of the first year of life the vast majority of lesions will have ceased proliferation and will be in early involution.

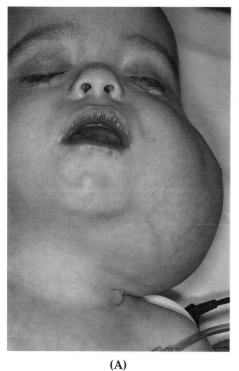

(A)

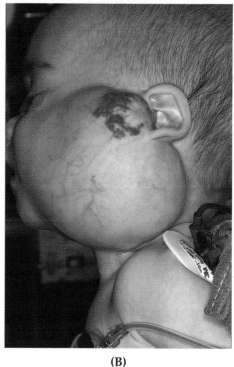

(B)

(C)

Figure **9.14**
(A, B) A 12-month-old infant with a massive parotid hemangioma. The child had been treated with steroids and interferon and was still in severe high-output failure. **(C)** The child 3 months after resection of the lesion. The child is no longer in cardiac failure, and all branches of the facial nerve are recovering.

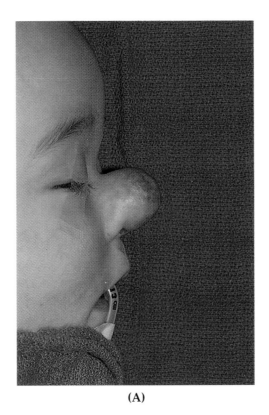

(A)

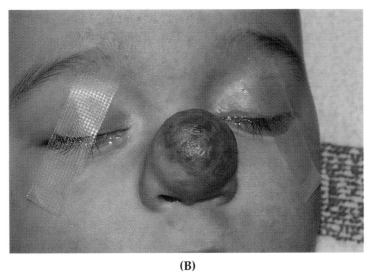

(B)

Figure **9.15**
(A, B) A large disfiguring nasal tip hemangioma.

9.4.4. Early Involution

Apart from complications and lesions involving certain anatomical sites, early involution is a time during which conservative management is appropriate. One can afford to let nature run its course until 3–4 years of age. This will allow a more accurate estimation of how much involution can be expected and how soon it is likely to be completed. Complete involution by age 6 years (rapid involution) will, in 60% of cases, be complete and need no further treatment. Should this not be the case (i.e., slow involution, not complete by 6 years) then corrective surgery will be necessary in 80% of cases. A further consideration should be the child's entry into school. If it becomes likely that corrective surgery will be needed, then it should, in ideal circumstances, be completed prior to this. Furthermore, because in so many cases multiple procedures are necessary, sufficient time should be allocated. Further consideration should also be given to psychosocial factors such as the child's apparent disposition and the parents' attitude. Clearly, then, the decision if and when to intervene is multifactorial.

Treatment should be directed at improving the child's appearance. Steroids are not effective against involuting hemangiomas and should therefore not be considered. Appropriate treatment modalities include laser photocoagulation, surgical resection, or both. The depth and location of the lesion will determine which is best. As a rule, laser photocoagulation is effective in treating superficial lesions or the superficial component of a compound lesion. Surgical resection is the most effective way of dealing with a deep lesion or the deep component of a superficial lesion.

9.4.4.1. Superficial Hemangiomas

The ectatic vascular channels seen during involution are ideally suited for laser photocoagulation. These vessels may be large, small, or intermediate. Small and intermediate vessels respond well to treatment with a flashlamp pumped-dye laser and should be treated in the conventional way described in Chapter 10. The larger vessels are more resilient and require more thermal energy and/or a longer exposure time. The newer generation of flashlamp pumped-dye lasers may well be adequate for these vessels, and because the pulse width is variable, one laser may be sufficient for all calibers. Unfortunately, at this point a different laser is necessary to treat these larger vessels. The longer ex-

posure times available with copper bromide lasers and pulsed fre-
quency-doubled Nd:YAG lasers make either of these lasers ideal. Both
lasers emit pulsed light with exposure times of 1–50 msec. By keeping
the energy density at 12 J/cm^2 or less, one is able to selectively destroy
the larger vessels with little or no damage to the overlying skin.
Moreover, the slower heating seen with these lasers will result in vasoconstriction as opposed to the coagulation and vascular rupture seen
with flashlamp pumped-dye lasers. The end point is thus disappearance of the vessels or, with larger vessels, a narrower grayish ghost-like
vessel. For a complete description of the technique, see Chapter 10.

Lakes of superficial ectatic telangiectasias also respond well to
treatment in the same way. An end point of vasoconstriction, together
with thermal shrinkage of interstitial collagen, will shrink the lesion
and, after two or three treatments, render the superficial component invisible. After a variable number of treatments (determined by the thickness and response of the lesion) a flat, mainly erythematous residuum
made up of much smaller vessels will remain. At this stage, a flashlamp
pumped-dye laser is most suitable, and the lesion should be treated in
the conventional way. At least two treatments with an interval of 3
months is required for an optimal result.

9.4.4.2. Deep Hemangiomas

The depth of subcutaneous hemangiomas precludes lasers as an effective modality of treatment. Apfelberg (1995), however, has overcome
this exclusion by developing a technique whereby the thermal energy
generated by an Nd:YAG laser is used in a nonspecific way to coagulate deep lesions. A bare quartz fiber coupled to an Nd:YAG laser is inserted into the substance of the hemangioma. With a low energy density to avoid charring, the temperature of the lesion is raised to
60°–80°C. Several passes are made through the hemangioma in this
way to coagulate as much tissue as possible. While this work is still in
its infancy, the concept warrants further consideration. Once the extent
of coagulation around a fiber at a given energy density has been determined, perhaps a more scientific approach will be possible. A three-
dimensional reconstitution of an MRI scan with the sterotactic placement of laser fibers in a predetermined way would destroy a given
quantity of the hemangioma. Furthermore, because the T2-weighted
images are temperature sensitive, real-time monitoring may be possible. Successive treatments could then be used to debulk the lesion as

much as is desirable. This work is still experimental, and the reader is urged to consult the literature prior to considering this form of treatment.

Surgical resection remains the mainstay of treatment for subcutaneous hemangiomas (Fig. 9.16). A computerized tomographic (CT) scan with contrast or an MRI will delineate the full extent of the lesion and should always be done prior to surgery. A coagulation profile should only be considered if a coagulopathy is suspected, and the need for blood typing and cross-matching should be determined by the size of the lesion and the past experience of the surgeon. During surgery, meticulous attention to hemostasis, facilitated by the use of a ther-

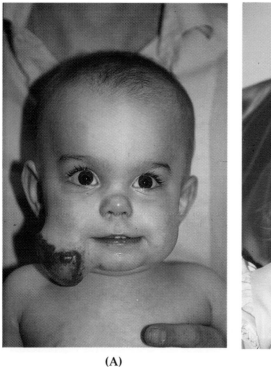 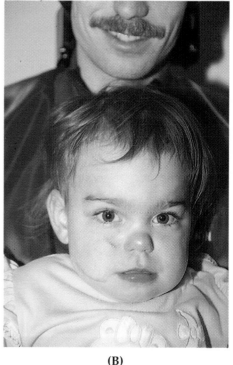

(A)   (B)

Figure 9.16
(A) An infant with a pedunculated facial hemangioma. This lesion is likely to eventually require surgical resection. At this stage it behaves like a tissue expander, thereby facilitating primary closure. **(B)** The same child about 6 months after resection of the lesion. The scar was revised shortly after this photograph had been taken and the child has done well.

moscalpel or contact laser surgery, will often diminish the need for a transfusion. However, if this is needed, in most cases donor-directed blood is preferable.

Because the object of surgical resection is to improve the appearance and/or function of the child, the need for complete resection, as opposed to debulking or partial resection, should be carefully considered and weighed against potential risks. If the lesion under consideration is an involuting hemangioma, allowances for this should be made. During involution, hemangiomas are replaced by fibro-fatty tissue. Overzealous resection without due consideration of this may overcorrect the problem and need revision at a later stage. It may therefore be prudent to leave a residuum, knowning full well that this will be replaced by fibro-fatty tissue once involution is complete.

Throughout this process, one should never lose sight of the natural history of hemangiomas. Treatment should never give a worse result than that seen with natural involution. Surgery should therefore err on the conservative side, and all attempts should be made to preserve normal structures. The surgical incision should follow relaxed tension lines if this is possible and should conform with the standards practiced in cosmetic surgery. A cosmetically appropriate scar should be the desired result (Figs. 9.17, 9.18, and 9.19).

9.4.4.3. Compound Hemangiomas

The approach to a compound lesion should include all of the principles previously discussed. However, in planning treatment, the cutaneous component should always be dealt with first. Prior photocoagulation of the surface will destroy some of the cutaneous hemangioma and replace it with fibrous tissue. This will enable the surgeon to incise through the skin and raise it as a flap. The fibrous tissue layer formed by photocoagulaton is a convenient plane to work in because destruction of vascular tissue will facilitate hemostasis. Also, it stands to reason that more skin will be preserved, which can later be used in reconstruction. This technique has been successfully used in a large series of cases (Waner and Suen, 1992) (Figs. 9.18, 9.20).

9.4.5. Late Involution

During late involution, most of the vascular channels disappear. A few do, however, persist as dilated capillary-like vessels and are

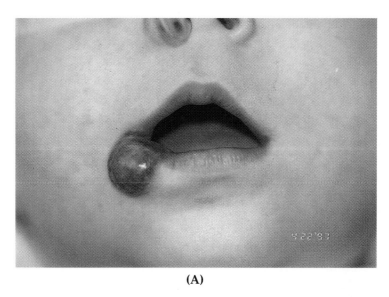

(A)

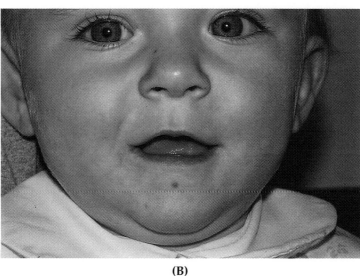

(B)

Figure 9.17
(A) A proliferating oral commisure hemangioma on a 6 month old. (B) Three months after resection. The scar conforms with the relaxed tension lines and thus is less noticeable.

made up of a single layer of endothelial cells on a basement membrane. At the same time, as the number of vascular channels diminishes, fibro-fatty tissue appears in the interstitium and replaces some of the vascular tissue. A variable amount of this tissue may persist after involution is complete. Another feature of involution is

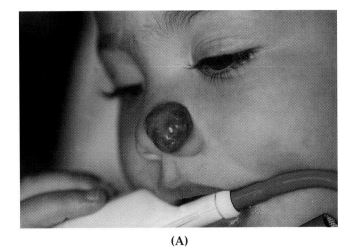

(A)

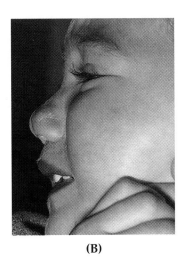

(B)

Figure **9.18**
(**A**) A proliferating lesion of the alar groove. (**B**) The child was treated with a pulsed-dye laser on two occasions at 6-week intervals preoperatively and then underwent surgical resection. The scar (in the alar groove) is barely visible.

epidermal atrophy, caused by an absence of the papillary dermis (which was replaced by the hemangioma during proliferation), stretching of the overlying skin by the hemangioma, and possibly elastolysis.

At this late stage in the hemangioma's life cycle, one should use the most appropriate means to improve the child's appearance and, in so

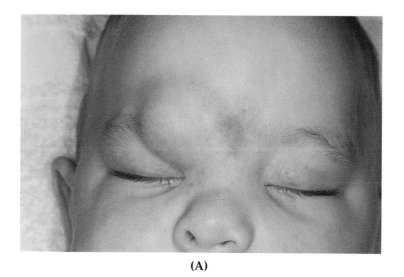

(A)

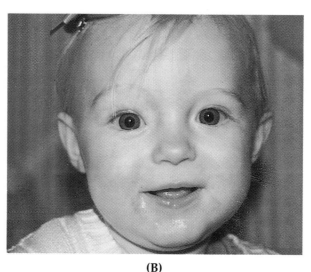

(B)

Figure **9.19**
(A) A 3-month-old infant with a rapidly proliferating deep hemangioma of the upper eyelid. The child was found to have early astigmatism. **(B)** Rather than systemic or intralesional steroids, the parents requested surgical excision. At 3 months postsurgery, the scar conforms well with the brow line and is hardly visible.

doing, his or her self-esteem. The problems that need attention are one or more of the following: telangiectasia, epidermal atrophy, and a noticeable mass of fibro-fatty tissue. (More often than not, the child will be of an age when peer group influence is strong, and no amount of counseling can completely neutralize this problem.)

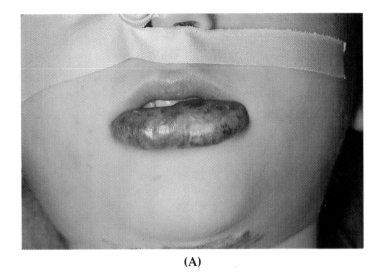

(A)

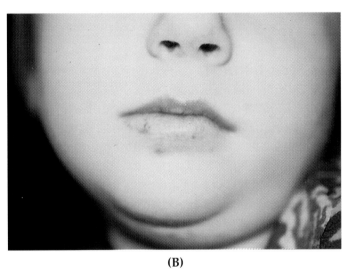

(B)

Figure **9.20**
(A) An 18-month-old child with a complete lower lip involvement. **(B)** The same child after surgical debulking and shortening through a wedge resection. The vermilion was rotated backward by means of a horizontal mucosal wedge excision carried through to the orbicularis oris.

9.4.5.1. Telangiectasia

An arborizing network of ectatic cutaneous capillary-like vessels is unsightly and can easily be remedied. These are either large or medium sized, in which case discrete telangiectasia, or small vessels, will be visible as an erythematous blush. Once again, the medium-sized and the larger vessels are best treated with a copper bromide laser or a pulsed KTP laser. Using adequate eye protection, each individual vessel should be traced out with a 1-mm focused beam. An end point of vaso-constriction should be achieved with little or no blanching of the over-lying skin. This will be seen as either complete disappearance of the vessel or a gray shadow of the vessel will remain. Complete healing takes place within 2 weeks, and the process may need to be repeated twice or even three times before complete resolution of all the vessels can be expected.

Alternatively, a red blush may be left with or without any dis-cernible telangiectatic vessels. Because the blush is made up of much smaller vessels, this is best treated with a flashlamp pumped-dye laser (Fig. 9.21). A 5-mm spot size is used, and, with an energy density of 6.5–7.5 J/cm^2, a uniform bluish gray discoloration will be produced. Any remaining larger vessels can then be treated with a copper bro-mide laser or a pulsed KTP laser in the usual manner.

9.4.5.2. *Residual Fibro-Fatty Tissue*
Residual fibro-fatty tissue may result in a visible mass, and this should be excised if it can be accomplished safely and with minimal scarring. Several residual ectatic vessels will make the tissue somewhat more vascular than expected. The surgeon should, therefore, exercise caution during these procedures and pay particular attention to hemostasis. Contact Nd:YAG laser surgery or a thermoscalpel and bipolar electro-cautery are most useful for this procedure. Because involution may be a slow process and may continue for up to 12 years, one should err on the conservative side. Overzealous contouring, together with contin-ued natural involution, could result in a defect. Allowances must there-fore be made for some spontaneous involution if the child is younger than 12 years of age. The age-old dictum "first do no harm" should al-ways be adhered to.

9.4.5.3. *Epidermal Atrophy*
Thin atrophic skin usually co-exists with telangiectasia and residual fi-bro-fatty tissue, and it is difficult to rectify. Until recently, dermabrasion was the only alternative and offered only marginal improvement. The associated risks were considerable and included hypertrophic scarring, hypopigmentation, and postinflammatory hyperpigmentation. The

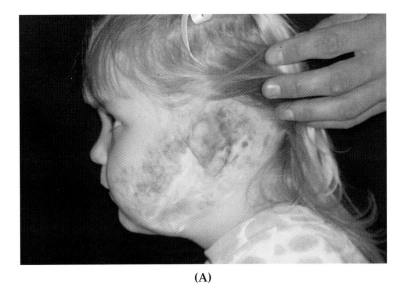

(A)

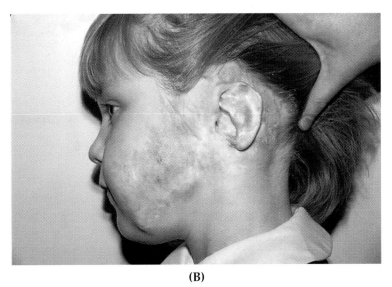

(B)

Figure 9.21
(A) A child with a parotid hemangioma in late involution. The child is left with residual telangiectasia, atrophic and hypertrophic scarring, and residual fibro-fatty tissue. (B) After excision of the atrophic/hypertrophic scarring, flashlamp pumped-dye laser treatment of the telangiectasia, and an otoplasty, the child's appearance has improved.

likelihood of these complications was in the region of 10%. Skin resurfacing with a CO_2 laser or an Er:YAG laser appears to be extremely promising. Recent work using this technique to treat atrophic scarring in children with complete or partial involution of the hemangiomas reported significant improvement in almost all patients. Potential risks were similar, but their incidence appears to be considerably less than that seen with dermabrasion (Waner, 1996).

With a modified pulsed or a flashscanned continuous wave CO_2 laser, the epidermis is vaporized to the level of the dermoepidermal junction. The papillary dermis is then treated in the same way but at this level, the dominant effect is heat shrinking of the collagen layer. As

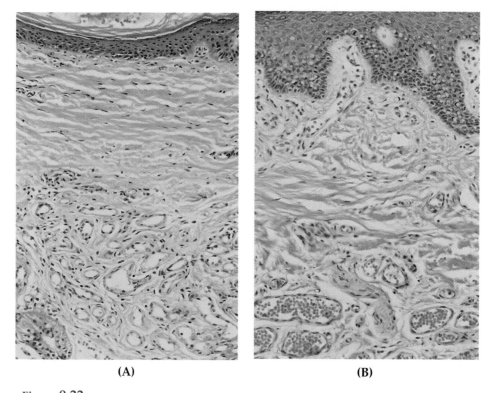

(A)   (B)

Figure 9.22
(A) The skin of a $2\frac{1}{2}$-year-old child with an involuting hemangioma. Note the atrophic changes (epidermal thinning and a loss of rete pegs). An atrophic scar is seen in the center of the field, but adnexial structures are still present despite this. (B) The same child 6 months after resurfacing. Note the appearance of normal rete pegs and a normal papillary dermis. A residual deep hemangioma remains in the reticular dermis and subcutaneous fat layer.

the thermal energy vaporizes water from the interstitium, the thermal effect on the surrounding collagen layer changes its quaternary structure and reduces its length. This will effectively smooth out the irregular scarred surface and, over the ensuing 6 months, thicken the papillary layer (Fig. 9.22). The atrophy will thereby by eliminated.

9.5. HEMANGIOMAS REQUIRING SPECIAL CONSIDERATION

Although the principles of management just outlined can be applied to all head and neck hemangiomas, lesions in certain anatomical locations warrant further consideration.

9.5.1. Hemangiomas of the Upper and Lower Lips

Both the upper and lower lips are commonly involved, but are also frequently associated with complications peculiar to this site. Lip hemangiomas frequently ulcerate, and bulky lesions often lengthen the involved lip. Ulceration is usually chronic, and an ulcerated lesion almost always heals with scar tissue formation that will ultimately need to be excised (Fig. 9.5). An increase in lip length usually persists, and this will also ultimately need correction. The timing of correction is crucial. In some instances earlier rather than later correction is preferred, whereas in other cases the opposite is true.

9.5.1.1. Lip Lengthening
Both upper and lower lip hemangiomas frequently stretch the lip, and a simple through-and-through wedge resection of the lip involved with hemangiomas will often correct this (Figs. 9.23, 9.24, 9.25). The desired length of the lip can be estimated by measuring the length of the opposite lip along the red margin. The length of tissue that can be resected from a lower lip should ideally leave it is the same length as the upper lip and vise versa. At worse, up to 20% of the length of the lower lip can be resected over and above the ideal length and still leave an aesthetically acceptable result. The axis of the incision should always follow the relaxed tension line, and it is imperative to realign the red margin. The best way to accomplish this is with prior surgical marking.

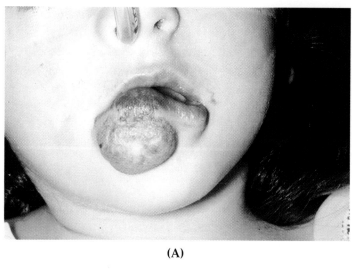

(A)

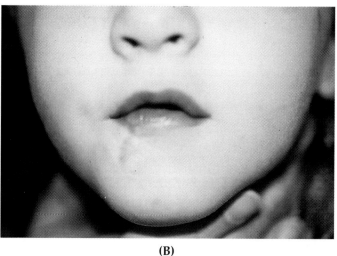

(B)

Figure 9.23
(A) An 18-month-old infant with a large full-thickness lower lip hemangioma. The lesion has elongated the lip by 20%–30%. **(B)** The same patient a year later. The child's mother was offered a scar revision but declined. She was pleased with the result.

9.5.1.2. *Upper Lip Lesions*

Hemangiomas of the lip may be subcutaneous (mostly in front of orbicularis oris), submucosal (behind oribcularis oris), or through and through. Unfortunately, full thickness lesions are common, and the orbicularis oris muscle is frequently involved. It is extremely difficult to

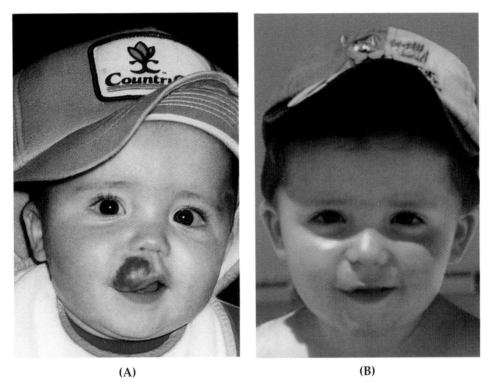

(A)   (B)

Figure **9.24**
(A) An upper lip hemangioma in an infant. The lesion has stretched the upper lip, resulting in lengthening of the lip that is unlikely to resolve. **(B)** A child 3 months after resection of the lesion. A residuum of superficial hemangioma, seen prior to surgery, still remains and will resolve with one or two treatments with a pulsed dye laser.

separate the hemangioma from muscle. In addition, the lip is usually lengthened, especially in the presence of a large, bulky hemangioma. A wedge resection of the hemangioma and lip will correct the lengthening unilaterally. An estimate of the amount of vermilion one can resect can be made by measuring the distance between the tubercle and the commissure on either side and then subtracting these lengths. The difference will indicate the excessive lengthening one has to deal with. Again, the axis of the incision should be parallel to the relaxed tension lines, and accurate realignment of the vermilion is essential.

9.5.1.3. *Lip Eversion*
Eversion of the vermilion is frequently seen with hemangiomas of most or all of the lower lip. The lip can be rotated back to a normal position

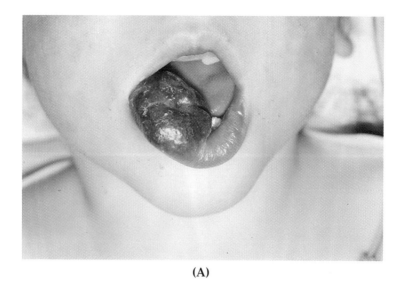

(A)

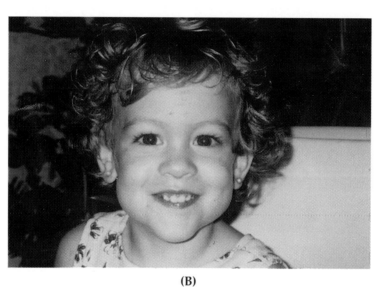

(B)

Figure 9.25
(A) An infant with a large ulcerated hemangioma of the lower lip. (B) The same
child 3 months after a wedge excision. The lesion had distended the lip (as it so often does), necessitating a wedge resection.

through a horizontal wedge resection of mucosa (Fig. 9.20). The width
of the mucosal strip that would need to be excised can be determined
by measuring the distance between the desired position and the uncor-
rected position. The wedge should be widest opposite the most everted
area (usually the midline) and taper on either side of this. It is usually

necessary to extend the wedge all the way to the level of the commissure on either side. The apex of the wedge should be just superficial to the orbicularis oris muscle, and the wound should be enclosed in two layers.

9.5.1.4. Lip Inversion
Lip inversion can usually be corrected with a dermal implant using *lyiophilized* dermis or dermis harvested from some other anatomical site. One can augment and evert both the lower and the upper lip to a more desirable position. Two 2-mm vermilion incisions are made 3–4 mm in from the commissure of the lip and parallel to the relaxed tension lines. With blunt dissection, a subcutaneous tunnel is developed between these two sites. The width of the tunnel should equal the width of the vermilion, and, using a tendon passer or any other suitable instrument, the dermal implant should be placed along this line. The wounds are closed with resorbable chromic sutures.

9.5.1.5. Hypertrophic Scarring
Hypertrophic scarring involving the vermilion will cause disfigurement. A linear strategically placed scar is preferable and less noticeable. The scar can therefore be resected through a wedge resection especially if a small or limited area is involved (Fig. 9.26). If the area is too extensive for surgical revision, skin resurfacing will allow the surgeon to "plane down the scar," and this will improve its appearance (Fig. 9.27). When using a skin resurfacing technique, it is important to match the scan or spot size with the width of the scar. In this way, the scar can be planed down to an appropriate level without affecting the surrounding skin.

9.5.1.6. Debulking Procedures
It may become necessary to debulk the lip, especially the upper lip, to improve the appearance of the child (Fig. 9.28). The procedure will depend on the extent and location of the hemangioma. A useful technique for debulking vermilion lesions is to approach the lesion from the mucosal surface. A horizontal incision made through the mucosa will enable the surgeon to elevate a mucosal flap as far as the red margin. The spongy hemangiomatous tissue can then be resected to leave a more normal-appearing vermilion.

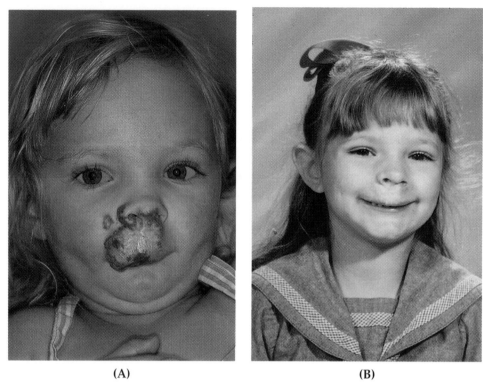

(A)   (B)

Figure 9.26
(A) A 2-year-old child prior to correction of a scarred upper lip with considerable residual hemangioma. **(B)** The residuum and scar were excised utilizing a vermiliocutaneous approach. The lip was then treated with a pulsed-dye laser several times and then skin resurfacing. Six months has elapsed since skin resurfacing.

9.5.2. Nasal Tip Hemangiomas

Hemangiomas of the nasal tip are more often than not deep hemangiomas and occupy the space between the skin of the nasal tip and the lower lateral cartilages. As the hemangioma expands, it distracts the cartilages, and this results in a bulbous nasal tip, the so-called "Cyrano" nose. Unfortunately, the hemangioma can become quite large and disfiguring.

A useful approach to these lesions is through an external rhinoplasty incision. The hemangioma will become obvious as soon as the surgeon enters the subcutaneous plane. A very thin, delicate skin flap should be elevated until the top of the hemangioma is reached. The surgeon should

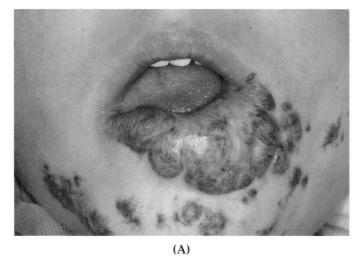

(A)

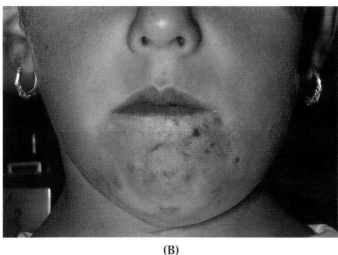

(B)

Figure 9.27
(A) An 18-month-old child with scarring of the lip as a consequence of ulceration.
(B) The scarred area was resected together with the residual hemangioma. The residuum was then treated with a flashlamp pumped-dye laser, and finally resurfaced with a CO_2 laser.

exercise extreme caution with this flap as it is very easy to "button hole" the skin. The hemangioma can then be dissected off the front of the nasal septum and between the lower lateral cartilages. Care should be exercised in avoiding damage to the cartilages, as this will likely affect subsequent carilagenous growth. Once the hemangioma has been removed, it is usu-

 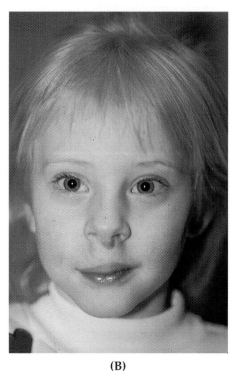

(A)   (B)

Figure 9.28
(A) A 3-year-old child with an extensive upper lip hemangioma. **(B)** The same patient 3 months after surgical debulking. An Abbe flap will be used to reconstruct a
philtrum when the child is old enough.

ally necessary to medialize the cartilages with one or two vicryl sutures
(Fig. 9.29 C,D). It is very important to place each of the sutures symmetri-
cally through the same area of each of the lower lateral cartilages to avoid
asymmetry of the nasal tip. The skin is then draped back over the wound,
and invariably excessive skin will be left; once the hemangioma has been
excised this can be trimmed. Because postoperative bleeding is a potential
risk, once the wound has been closed Steri strips should be placed over
the nasal tip to lightly compress the skin against the subcutaneous tissues
and obliterate the potential dead space (Fig. 9.29E). Any excessive skin
will be taken up over the next 6–8 weeks, and any redundant skin after
that can be reduced with a resurfacing technique. Although there are no
data to substantiate this, more clinicians agree that nasal tip hemangiomas
involute slowly; it is therefore probably best to remove these hemangiomas sooner rather than later. Another advantage of earlier removal is
that the surgical plane is more distinct and easier to access.

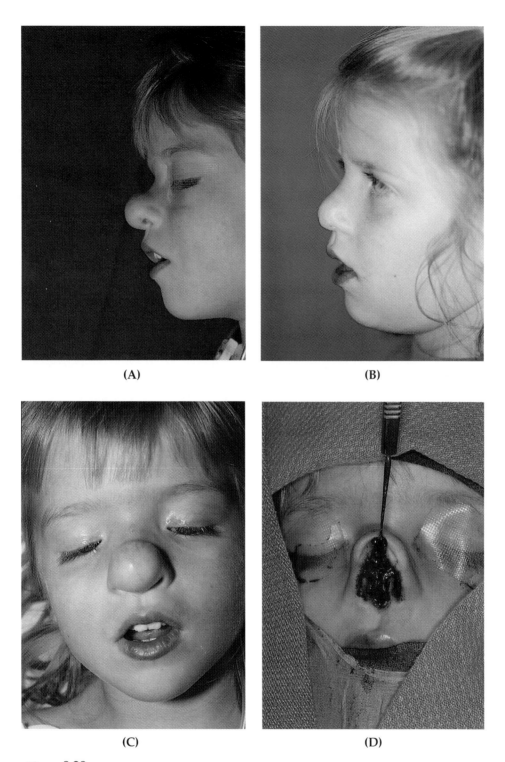

(A) (B)

(C) (D)

Figure 9.29
(A) A 3-year-old child with a slow involuting nasal tip hemangioma. **(B)** The same child, 18 months status postsurgical resection. **(C)** An anterior view of the same patient preoperatively. **(D)** An open rhinoplasty technique was used and the lower lateral cartilages were re-approximated with an absorbable suture.

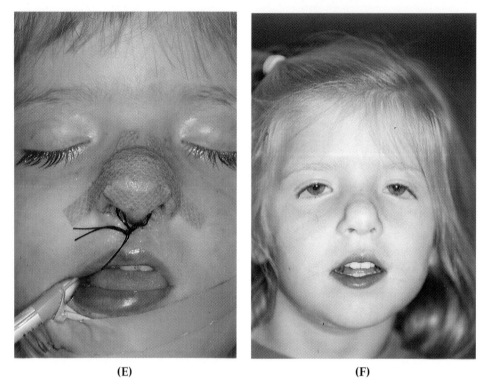

(E)   (F)

Figure **9.29** *continued*
(E) Postoperatively, compression with Steri strips is necessary to prevent a postoperative hematoma. **(F)** An anterior view of the patient 18 months after resection of the hemangioma.

9.5.3. Parotid Hemangiomas

One of the peculiarities of parotid hemangiomas is that they almost always involve the entire parotid gland. The hemangioma seems to follow the anatomy of the parotid and extend throughout the course of the gland. Occasionally, the overlying skin is involved, but more often than not this is only partially involved or spared. Although parotid hemangiomas appear to invoute well, a giant parotid hemangioma occasionally warrants resection. Apart from being extremely disfiguring, these lesions may occasionally precipitate high-output failure (Fig. 9.14). Parotid hemangiomas are also occasionally implicated in Kasabach-Merritt syndrome, and, although the primary management of Kasabach-Merritt syndrome is pharmacological, when faced with a potentially fatal outcome in a patient who is refractory to treatment the surgeon should be prepared to remove or debulk the lesion as a last resort.

Parotid hemangiomas are the most surgically challenging and should only be attempted by an experienced surgeon familiar with conventional parotid surgery. Because the hemangioma will ultimately involute, it is unnecessary to remove the entire lesion. Removal of the superficial lobe of the parotid is usually all that is necessary to correct the disfigurement and/or relieve the high-output failure, and this should be accomplished with facial nerve preservation. After the parotid fascia has been located and a prefascial skin flap elevated to the front of the parotid gland, the facial nerve should be located either behind the parotid gland or in front of it by following the usual anatomical landmarks. We prefer to locate the main trunk of the facial nerve and then trace it forward through the gland. However, should this not be possible, the surgeon may elect to locate one or more of the peripheral branches and then follow them back through the substance of the gland to the main trunk. One of the most difficult aspects of this procedure is the maintenance of hemostasis. Careful, slow dissection with a thermoscalpel and bipolar electrocautery is necessary. It is not unusual for this part of the dissection to take many hours. All bleeding should be controlled as soon as it is encountered and not at some later stage because the surgeon may become overwhelmed and lose control of the facial nerve. Once the main trunk and several branches have been located, a plane should be declared and all the hemangioma superficial to this plane should be removed. Some large vessels may be encountered during this part of the dissection. These should once again be controlled as soon as they are encountered. This is one of the few occasions where it may be necessary to transfuse the child. Careful preoperative planning will ensure that donor-directed blood is available, or, if this is not possible, at least 2 units of whole blood should have been typed and cross-matched. (See Fig. 9.14.)

9.5.4. Subglottic Hemangiomas

The insidious onset of symptoms of an upper airway obstruction after the first 6–8 weeks of life is highly suggestive of a subglottic hemangioma. The most common symptoms include inspiratory or *biphasic* stridor exacerbated by crying or an upper respiratory tract infection. The subglottis is a site of *predilection*, and, as with other sites, this represents an area of fusion between the bronchial and the postbronchial or somatic structures. The most common site within the subglottis appears to be the posterolateral aspect, and for some hitherto unexplained reason the left side appears to be more common (Brodsky et al., 1983). Occasionally, a bilateral or a circumferential lesion is found. Subglottic hemangiomas are usually submucosal, and their appearance is not dissimilar from a submucosal hemangioma found anywhere else in the aerodigestive tract. It is usually a bright red and firm lesion during active proliferation and a dull, dusky maroon color during the quiescent phase or during involution. While proliferating lesions are usually firm, involuting lesions are soft and, during late involution, tend to be compressible. Occasionally, the lesion may be part of a large peritracheal hemangioma in which case the submucosal or tracheal presentation may merely be the tip of an iceberg. In these instances, the lesion is pale and is separated from the lumen of the trachea by the full thickness wall of the trachea.

The diagnosis is usually made at endoscopy, and occasionally an MRI will be helpful, especially in the presence of an atypical lesion. Simpler radiographic procedures may also be helpful. A soft tissue penetration anteroposterior film may demonstrate asymmetrical narrowing of the subglottis, and an esophagogram will exclude a vascular ring as a possible cause of the symptoms. A biopsy should only rarely be necessary to establish the diagnosis.

Treatment options include expectant, Pharmacotherapy (steroids, interferon) and surgical ablation (CO_2 laser ablation, open resection, or tracheostomy). The aim of treatment is to maintain or establish a safe airway. Because the disease is self-limiting and prevention is better than cure, we recommend aggressive measures early in the disease process to obviate the need for a tracheostomy.

Children with small lesions of less than 20% compromise can be treated expectantly and require careful follow up. Once the child becomes symptomatic, pharmacotherapy should be the first line of treatment. An adequate dose of prednisone or prednisolone will in most cases shrink the lesion and protect the airway. We recommend a dose of 5 mg per kg body weight per day for 2–3 weeks followed by a slow taper over 2–3 months. With careful clinical follow up, repeat endoscopy should not be necessary. Rebound growth of the hemangioma during the tapering dose can be effectively controlled by increasing the dose back up to a maximum for 1–2 weeks and then commencing a further taper. This can be done until one no longer experiences rebound growth on the taper. With prolonged treatment, the child will develop cushingoid features and several of the effects of long-term high-dose steroids; however, these all appear to be reversible after cessation of steroids (Sadan and Wolach, 1996).

Failure to control the airway in this manner will necessitate surgical intervention. A well-localized lesion can be ablated endoscopically with a CO_2 laser (Fig. 9.30). A single attempt or occasionally a second attempt can be made (Healy et al., 1990, 1994). The surgeon should re-

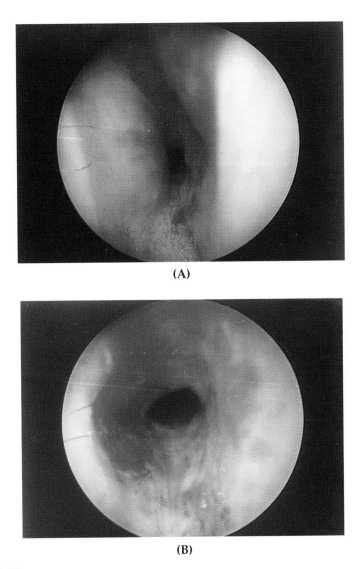

(A)

(B)

Figure 9.30.
(A) A typical unilateral subglottic hemangioma. (B) Immediately after CO_2 laser ablation

strict himself to ablation of the hemangioma and avoid damaging the cartilagenous skeleton of the larynx. A deeper lesion, a diffuse lesion, or alternatively a circumferential lesion is best managed by performing a tracheostomy. CO_2 laser ablation of these lesions is associated with an increased risk of stenosis and is therefore best avoided. If the surgeon does, however, elect to proceed with laser ablation of a circumferential lesion, this should at least be staged.

It is imperative that the surgeon bear in mind the natural history of hemangiomas. Laser ablation during the first year of life will invariably result in some degree of rebound growth unless the lesion has been completely ablated. The same is true of steroids. However, during involution, steroids will no longer be effective and incomplete laser ablation will no longer be followed by rebound growth. Failure to control the lesion and/or maintenance of an adequate airway in this way should be followed with a tracheostomy.

Other modalities that have gained popularity include open surgical resection, interferon-α and intralesional steroid injections. Open surgical resection through a cricothyrotomy has gained some support (Mulder and van den Broek, 1989; Seid et al., 1991). The surgeon should exercise caution in preselecting appropriate cases. Once again, isolated unilateral lesions are more easily resected without complications. Because this approach is relatively new, the surgeon should exercise caution when selecting this modality.

Initial success with interferon-α has been tempered by the disturbing occurrence of spastic dyplegia in a significant number of children (Ezekowitz et al., 1992; Ohlms et al., 1994; Soumekh et al., 1996). The first case of spastic dyplegia was reported in a child on interferon-α for laryngeal papilloma in 1988 (Vesikari et al., 1988); since then several cases have been reported (Mulliken et al., 1995). Interferon should therefore only be used in life-threatening situations and then only if there is no other alternative. This therefore precludes its use in airway management until the issue of spastic dyplegia has been resolved.

Lastly, the use of intralesional steroids has also gained some support (Meeuwis et al., 1990). Again, bearing in mind the natural history of hemangiomas, more than one injection may be needed. Intralesional steroids were initially used to avoid their systemic effects, but recent reports have implicated intralesional steroids in several of the systemic side effects. It would, however, appear that the incidence with intralesional steroids is less than that with systemic steroids. In the context of subglottic hemangiomas, the child should be anesthetized and the steroid carefully injected into the substance of the hemangioma. Thereafter, the child may need to be kept intubated for at least a week to 10 days postinjection. The use of intralesional steroids is likely to decrease the incidence of systemic side effects and appears to be less traumatic than superficial ablation with a CO_2 laser or open surgical resection. The necessity of intubation in the immediate postinjection period has discouraged more widespread use of this form of treatment.

9.6. THE MANAGEMENT OF COMPLICATIONS

9.6.1. Ulceration

Ulceration is the most frequent complication and occurs in about 5% of all lesions. (Margileth and Museles, 1965). For some reason, it is more common in certain sites such as the upper lip and the anogenital area. Ulceration usually occurs during periods of rapid proliferation and, left untreated, may persist throughout the duration of this phase. The mechanism is thought to be related to the rapid growth and distension of the overlying skin. However, one often see ulceration in small flat lesions. It is thus likely that either tumor necrosis factor or some similar factor released by the mast cells is responsible. Concerning the duration of ulceration, two factors are thought to be determinants: the continued growth of the hemangioma and sepsis. Our objectives should therefore be to prevent and/or control sepsis and, if possible, to slow proliferation until after the ulcer has healed. Routine wound care as well as the regular use of topical antibiotics such as polysporin and bacitracin will provide a moist environment and facilitate moist wound healing. In addition, recent evidence suggests that treatment with a flashlamp pumped-dye laser will slow or retard proliferation and by so doing afford the epidermis an opportunity to heal over the defect (Morelli et al., 1991; Garden et al., 1992). The wound should be carefully cleansed prior to treatment to remove any soft escar that might have accumulated. A hard escar, on the other hand, is best left because any attempt at removal will result in bleeding. It is important to avoid bleeding because blood will prevent the penetration of laser light, therefore negating the purpose of treatment. Using a standard 5-mm spot size, the entire lesion should be treated at 6.5–7.5 J/cm^2 to an end point of bluish black discoloration. Treatment should extend up to and include the margin of the ulcer and as any other area of superficial hemangioma. It is possible that treatment will rupture one or more small vessels and result in intraoperative hemorrhage. With a cotton-tipped applicator, the point of bleeding should be located and controlled with firm pressure while treatment is completed. A repeat treatment may be necessary, and this should be carried out 4–6 weeks later. In most cases, only one or two treatments are all that is required to effect healing.

Laser treatment appears to be more successful with flat lesions that have ulcerated and should be avoided with thick proliferating lesions since it is likely to increase the size and depth of the ulcer. Laser treatment is probably also best avoided while the patient is on steroids because this may affect the patient's ability to heal. It seems a paradox that treatment with a laser, itself a destructive process, is able to promote healing. The probable answer lies in its ability to destroy hemangioma in the base of the ulcer as well as in its immediate vicinity, thereby allowing the advancing edge of healing epithelium to cover the defect. Lastly, it is important to realize that ulceration almost invariably results in the formation of scar tissue, which will persist as either a hypertrophic or an atrophic scar even after the hemangioma has involuted. This will clearly have an impact on the way the patient is managed because the result of natural involution is bound to be poor and require corrective surgery. For this reason, we have adopted a more aggressive approach with lesions that have ulcerated and advocate early intervention.

9.6.2. High-Output Cardiac Failure

High-output cardiac failure is best managed by a pediatric cardiologist. When faced with intractable failure during the proliferative phase, aggressive pharmacotherapy should be tried. If this is unsuccessful, it may be necessary to consider surgical intervention and/or embolization. Once the lesion has ceased proliferating, surgical intervention and/or embolization may be necessary. Interferon may be used when faced with life-threatening failure, and after all other alternatives have been exhausted.

9.6.3. Kasabach-Merritt Syndrome

The management of Kasabach-Merritt syndrome is extremely challenging. Unfortunately, no single theraputic modality is universally successful, and the mortality rate, even with appropriate management, is believed to be in the range of 50% (Hatley et al., 1993). The striking clinical and histological differences between Kasabach-Merritt syndrome and classic hemangiomas have led some to believe that these cases are not a complication of hemangiomas but rather part of a separate entity (Enjolras, 1997). A review of several cases revealed histological features consistant with kaposiform hemangioendothelioma and tufed angioma. Whether this is true of all cases of Kasabach-Merritt syndrome remains to be seen. Another confusing issue concerns the coexistence of consumptive coagulopathy together with the typical thrombocytopenia of Kasabach-Merritt syndrome.

The presence of dark purple, markedly edematous skin overlying the lesion, petechial and ecchymotic hemorrhages, and prolonged bleeding are highly suggestive of Kasabach-Merritt syndrome. Baseline investigations should include a blood count, serum fibrinogen, fibrin split products, prothrombin time, and partial thromboplastin time. Uncomplicated Kasabach-Merritt syndrome results in thrombocytopenia. Occasionally, a consumptive coagulopathy complicates Kasabach-Merritt syndrome, and typical findings include a decreased serum fibrinogen and increased serum fibrin split products. The management of uncomplicated Kasabach-Merritt syndrome should be medical:

- In the first instance, a large dose of prednisone or prednisolone (5 mg/kg body weight) should be given. A response as evidenced by an increase in the platelet count and a decrease in the degree of edema should prompt the physician to continue this level of steroids for at least 4 weeks and then slowly taper over a further 2 months. While the child is still on seroids, he or she should be monitored closely, and any intercurrent infection or stress should be treated vigorously and covered with an increased dose of steroids. The complete blood count should be monitored daily until the platelet count is within the normal range and then twice weekly while the child is still on steroids.

- A failed or inadequate response may necessitate an even higher level of steroids (up to 10 mg/kg for a short interval), or alternatively, the physician may elect to start interferon-α. The urine β-fibroblast growth factor (β-FGF) level should be measured, as this will give an indication as to whether the lesion will respond. An empiric dose of 3×10^{-6} units/m^2 should be given by daily subcutaneous injection and the complete blood count monitored daily until the platelet count is normal. The child should remain on interferon for at least 6–8 months, and during this time the child should be monitored carefully for any side effects. In view of the recent report of spastic dyplegia and the uncertainty of its incidence, a baseline neurological examination as well as monthly assessments should be undertaken. Refer to Chapter 8 for a more complete discussion of interferon.

- If at all possible, one should avoid platelet and blood product transfusions, as these will be trapped in the lesion and may exacerbate the sequestration. Although there is no universally accepted threshold below which blood products should be transfused, there clearly is, and this should be determined on an individual basis. It may, however, be necessary to infuse red cell concentrate more readily if the hematocrit falls below 25%.

- A failed response to both steroids and interferon leaves one in the dubious position of having to consider surgical intervention or embolization. At this point we do not know whether embolization will have an adverse effect on the sequestration of platelets. With regard to surgery, the patient *should be* transfused preoperatively with fresh frozen plasma, platelets, and red cell concentrate (if the hematrocrit is less than 25%) to improve the coagulation profile. The surgeon should be prepared to encounter extensive intraoperative hemorrhage and should have sufficient blood available. This can be an extremely hazardous procedure.

In the presence of Kasabach-Merritt syndrome complicated by a consumptive coagulopathy, the patient should be treated in the exact same way. Again, we avoid transfusing blood products (unless it becomes essential), and we avoid heparin because it potentiates the effect of growth factors on endothelial cells in vitro and it mobilizes β-FGF, a potent angiogenic factor, from the extracellular matrix (Mulliken et al., 1995; Taylor and Folkman, 1982; Sudhalter et al., 1989; Folkman et al., 1988).

It seems likely that Kasabach-Merritt syndrome is a complication of a variant of kaposiform infantile hemangioendothelioma. The latter does exist as a separate entity and our own limited experience with this lesion, as well as that of others, has been unfavorable (Niedt et al., 1989; Tsang and Chan, 1991). All cases have been unresponsive to therapy.

REFERENCES

Apfelbery, D.B.: Intralesional laser photocoagulation: Steroids as an adjunct to surgery for massive hemangiomas and vascular malformations. Ann. Plast Surg. 35:144, 1995.

Bowers, R.E., Graham, E.A., and Tomlinson, K.: The natural history of the strawberry nevus. Arch. Dermatol. 82:667–680, 1960.

Brodsky, L., Yoshpe, N., and Rulaen, R.: Clinical–pathological conclates of congenital subglottic hemangiomas. Ann. Otol. Rhinol. Laryngol. (suppl.) 105: 92–108, 1983.

Cooper, A.G., and Bolande, R.P.: Multiple hemangiomas in an infant with cardiac hypertrophy. Pediatrics 35:27–33, 1965.

Dupont, C., Chabrolle, J.P., DeMontis, G., et al.: Angiome cutane et insuffisance cardiaque. Ann. Pediatr. (Paris) 24:37–42, 1977.

Enjolras, O., Wassef, M., Mozoyer, E., Frieden, I., Rieu, P., Drouet, L., Taieb, A., Stalder, J.F., and Escande, J.P.: Infants with Kasabach-Merritt Syndrome do not have "true" hemangiomas. J. Pediatrics 130, 631; 1997.

Ezekowitz, R., Phil, D., Mulliken, J., and Folkman, J.: Interferon alfa–2a therapy for life-threatening hemangioma of infancy. N. Engl. J. Med. 326:456, 1992.

Finn, M., Glowacki, J., and Mulliken, J.: Congenital vascular lesions: Clinical application of a new classification. J. Pediatr. Surg. 18:894, 1983.

Folkman, J., Klagsbrun, M., Sasse, J., Wadzinski, M., Ingber, D., and Viodavsky, I.: A heparin-binding angiogenic protein: Basic fibroblast growth factor is stored within basement membrane. Am. J. Pathol. 130:393–400, 1988.

Garden, J.M., Bakus, A.D., and Paller, A.S.: Treatment of cutaneous hemangiomas by the flashplump-pumped pulsed dye laser: Prospective analysis. J. Pediatr. 120:555–560, 1992.

Hatley, R.M., Sabio, H., Howell, C.G., Flickinger, F., and Parrish, R.A.: Successful management of an infant with a giant hemangioma of the retroperitoneum and Kasabach-Merritt syndrome with alpha-interferon. J. Pediatr. Surg. 28:1356–1357, 1993.

Healy, G.B., Fearon, B., French, R., et al.: Treatment of subglottic hemangioma with the carbon dioxide laser. Laryngoscope 90:809–813, 1990.

Healy, G., McGill, T., and Friedman, E.M.: Carbon dioxide laser in subglottic hemangioma: An update. Ann. Otol. Rhinol. Laryngol. 93:370–373, 1984.

Hidano, A., and Nakajima, S.: Earliest features of the strawberry birthmark in the newborn. Br. J. Dermatol. 87:138, 1972.

Larcher, V., Howard, E., and Mowat, A.: Hepatic hemangiomata, diagnosis and treatment. Arch. Dis. Child. 56:7–14, 1981.

Lister, W.A.: The natural history of strawberry naevi. Lancet 1:1429, 1938.

Margileth, A.M., and Museles, M.: Cutaneous hemangioma in children. J.A.M.A. 194:523–526, 1965.

Meeuwis, J., Bos, C.E., Hoeve, L.J., and van der Voort, E.: Subglottic hemangiomas in infants: treatment with intralesional corticosteroid injection and intubation. Int. J. Pediatr. Otohinolaryngol. 19:165, 1990.

Morelli, J.G., Tan, O.T., and Weston, W.L.: Treatment of ulcerated hemangioma with the pulsed tunable dye laser. Am. J. Dis. Child. 145:1062–1064, 1991.

Mulliken, J.B., Boon, L.B., Tokahashi, K., Ohlms, L.A., Folkman, J., and Ezekowitz, A.B.: Pharmacologic therapy for endangering hemangiomas. Curr. Opin. Dermatol. 109–113, 1995.

Mulder, J.J.S., and van den Broek, P.: Surgical treatment of infantile subglottic hemangiom. Int. J. Pediatr. Otorhinolarngol. 17:57, 1989.

Niedt, G.W., Greco, M.A., Wieczorek, R., Blanc, W.A., and Knowles, D.M.: Hemangioma with Kaposi's sarcoma like features: Report of two cases. Pediatr. Pathol. 9:567–575, 1989.

Ohlms, L., McGill, T., Jones, D., and Healy, G.: Interferon alfa-2A therapy for airway hemangiomas. Ann. Otol. Rhinol. Laryngol. 1:103, 1994.

Pereyra, R., Andrassy, R., and Mahour, G.H.: Management of massive hepatic hemangioma in infants and children: A review of 13 cases. Pediatrics 70:254–258, 1982.

Sadan, N., and Wolach, B.: Treatment of hemangiomas with high doses of prednisone. J. Pediatr. 128:141–146, 1996.

Seid, A.B., Pransky, S.M., and Kearns, D.B.: The open surgical approach to subglottic hemangioma. Int. J. Pediatr. Otorhinolaryngol. 22:85, 1991.

Sheu, H.M.N., Yu, H.S., and Chang, C.H.: Mast cell degranulation and elastolysis in the early stages of stria distensiae. J. Cutan. Pathol. 18:410–416, 1991.

Sudhalter, J., Folkman, J., Svahn, C., Bergendal, K., and D'Amore, P.: Importance of size, sulfation and anticoagulant activity in the potentiation of acidic fibroblast growth factor by heparin. J. Biol. Chem. 26:6892–6897, 1989.

Taylor, S., and Folkman, J.: Protamine is an inhibitor of angiogenesis. Nature 297:307–312, 1982.

Tsang, W.Y.M., and Chan, J.K.C.: Kaposi-like hemangioendothelioma: A distinctive vascular neoplasm of the retroperineum. Am. J. Surg. Pathol. 15:982–989, 1991.

Vaksman, G., Rey, C., Marache, P., et al: Severe congestive heart failure in newborns due to giant cutaneous hemangioma. Am. J. Cardiol. 60:392–394, 1987.

Vesikari, T., Nuutila, A., and Cantell, K.: Neurologic sequelae following inter-feron therapy of juvenile laryngeal papilloma. Acta Paediatr. Scand. 77:619–622, 1988.

Waner, M., and Suen, JY.: Treatment of hemangiomas of the head and neck. Laryngoscope 102:1123–1132, 1992.

Waner, M., Suen, J.Y., Dinehart, S., and Mallory, S.: Laser photocoagulation of superficial proliferating hemangiomas. J. Dermatol. Surg. Oncol. 20:1–4, 1994.

Waner, M.: Laser resurfacing and the treatment of involuting hemangiomas. Lasers Surg. Med. 40 (219) Suppl. 8, 1996.

Treatment Options for the Management of Vascular Malformations

MILTON WANER, M.D., F.C.S. (S.A.) AND JAMES Y. SUEN, M.D., F.A.C.S.

Because vascular malformations never involute, the rationale for their treatment needs no explanation. The choice of treatment modality will, however, depend on several factors. These include its vascular content, anatomical location, and depth. One should also bear in mind the natural history of these lesions and choose the most appropriate time for treatment. In general, the treatment modalities for the management of vascular malformations can be considered under four major categories: laser treatment, sclerotherapy, embolization, and surgical excision. A review of each of these modalities is important and helpful in understanding the management of vascular malformations.

Hemangiomas and Vascular Malformations of the Head and Neck, Edited by Waner, M.D. and Suen, M.D.
ISBN 0471-17597-8   © 1999 Wiley-Liss, Inc.

10.1. LASER PHOTOCOAGULATION

The possibility of using a laser to treat subcutaneous vessels through intact skin became a reality in the late 1980s (Anderson and Parrish, 1983; Tan et al., 1986; Morelli et al., 1986). Prior to this, lasers were used to destroy vascular lesions, but the process was nonselective. Adjacent skin or mucosa was invariably destroyed, and the risk of scarring was therefore much greater. In a sense, these earlier devices were little more than an expensive electrocuatery device. The advent of lasers capable of emitting pulsed yellow light and a greater understanding of the intricacies of light–tissue interaction changed everything. Using this form of light we were able to limit and eventually eliminate thermal destruction of all but the target vessels. This process has since become known as *selective photothermolysis.*

10.1.1. Selective Photothermolysis

Selective photothermolysis is, as its name implies, a process whereby ectatic blood vessels are selectively destroyed. The term *selective* implies that there is no damage to surrounding tissue. To do this, there are two prerequisites: (1) an appropriate wavelength and (2) an appropriate exposure time. These are interrelated, and both must be met for true selective photothermolysis.

10.1.1.1. *Appropriate Wavelength*

Hemoglobin is a convenient target chromophore because it is totally intravascular, abundant, and conveniently absorbs light in the visible spectrum (Fig. 10.1). Both hemoglobin and oxyhemoglobin exhibit similar absorption peaks. Oxyhemoglobin has three absorption peaks: α at approximately 415 nm (blue light); $\beta1$ at approximately 540 nm (green light); and $\beta2$ at approximately 580 nm (yellow light). Hemoglobin, on the other hand, has only two absorption peaks: α at approximately 430 nm (blue light); and β at approximately 550 nm (green light).

Light at 600 nm will therefore be absorbed by both hemoglobin and oxyhemoglobin, although hemoglobin will absorb this wavelength more strongly. With regard to port-wine stains, most of the vessels are in the upper papillary dermis, with the mean vessel depth being 0.365–0.665 m (Rosen, 1983). In the visible spectrum, the longer the wavelength of light, the deeper the penetration within tissue. At *500*

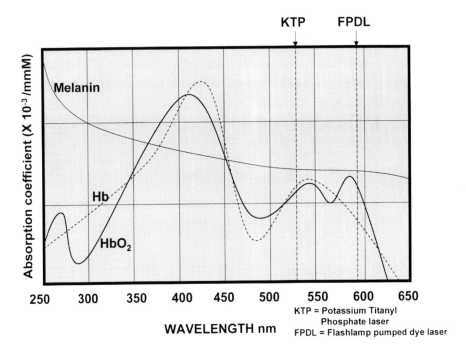

Figure 10.1
The absorption spectrum of hemaglobin (Hb), oxyhemaglobin (HbO$_2$), and melanin.

nm, only 50% of the incident light reaches a depth of 0.16 mm. At *600 nm*, 50% of the incident light will reach 0.38 mm, and 10% of the incident light will reach 1.27 mm.

Taking both the absorption spectrum of hemoglobin and the depth of penetration of visible light into account, it would seem that a laser emitting light at between 580–600 nm would be ideal. Several lasers are capable of producing this wavelength. These include the argon-ion pumped-dye laser, krypton laser, copper vapor laser, and flashlamp pumped-dye laser.

Both the Nd:YAG laser and its frequency doubled version the potassium titanyl phosphate (KTP) laser also emit wavelengths that are absorbed by oxyhemoglobin. The KTP laser at 532 nm emits light that corresponds with the β1 peak of oxyhemoglobin absorption and is close to the β peak of hemoglobin (Fig. 10.1). Despite the fact that there is ample absorption at this wavelength, the depth of penetration of this shorter wavelength is less and there is greater absorption by melanin. This laser is therefore probably less effective in treating deeper vascular lesions especially in darker pigmented patients (skin types III and IV), but there are advantages. The laser is a solid-state device and can be easily manipulated to produce ideal exposure times. This laser, therefore, enjoys some popularity and is especially useful in treating medium-sized vessel disorders.

The Nd:YAG laser at 1,064 nm emits light that is only slightly absorbed by hemoglobin, but the depth of penetration is much greater (approximately 5 mm). This laser is therefore useful for treating deeper vascular lesions such as venous malformations.

10.1.1.2. Exposure Time

The light energy absorbed by the target chromophore is converted to thermal energy, which then destroys the vessels. Regardless of the wavelength, nonselective damage will result if the exposure time is too long (Greenwald et al., 1981). Thermal energy will diffuse into surrounding tissues in spite of the light having been absorbed selectively by hemoglobin and oxyhemoglobin. To prevent this, the next prerequisite must be met—appropriate exposure time. The length of time a vessel is exposed to light is therefore of paramount importance. If energy is delivered to an object faster than it can diffuse away from that object, a high temperature gradient will exist between that object and its surrounding tissue, which will then result in spatial confinement of the thermal energy delivered by the laser. The time constant that determines this is known as the *thermal relaxation time,* and it measures the time required for the target to cool to 50% of its initial value (immediately after laser exposure) through heat transfer to its surroundings. It is thus a measure of the cooling time.

The thermal relaxation time is determined by the size and the optical characteristics of the target. For ectatic vessels commonly seen in a port-wine stain, this figure can vary between 1 and 10 msec (Fig. 10.2) (Dierickx et al., 1995). If the exposure time is equal to or less than the thermal relaxation time of the blood vessels, selective damage of the microvasculature will take place. Furthermore, because very small vessels are more efficient at dispersing heat, they are less likely to be damaged than the abnormal vessels, which happen to be much larger. Therefore, if the exposure time is approximately the same as or a little less than a particular vessel's thermal relaxation time, only large ectatic vessels that size will be destroyed. The smaller, normal vessels will be spared. This concept is important as it explains the reason why we are

• Vessel diameter	• Thermal relaxation time
30μ	0.86ms
40μ	1.54ms
50μ	2.40ms
100μ	9.60ms

Figure 10.2
The relationship between vessel diameter and thermal relaxation time (Dierickx et al., 1995).

able to destroy larger vessels and at the same time spare the smaller normal vessels. We refer to this phenomenon as *thermokinetic selectivity.*

Vascular malformations are made up of vessels that are ectatic; however, the degree of ectasia among the different types of lesions varies. Because the thermal relaxation time of a vessel is determined by its diameter and because the exposure time or pulse width of a particular laser is fixed, simply knowing these variables will allow one to predict which laser is best suited for each of the lesions (Fig. 10.2). To further explore this concept, we use the terms *small, medium,* or *large vessel disorders.*

Small vessel disorders include grades I, II, and III venular malformations (vessel diameter <100 μm). Medium vessel disorders include grade IV venular malformations and early venous malformations (vessel diameter between 100 and 400 μm). Large vessel disorders include grade V venular and venous malformations (vessel diameter >400 μm).

The exposure time of a laser will determine whether it is suitable for a particular lesion. In the field of cutaneous vascular ectasias, most of the work has been done with a flashlamp pumped-dye laser (Orten et al., 1996; Tan et al., 1989; Garden et al., 1988). The earlier version of this laser emitted a wavelength of 585 nm (yellow light) and a pulse duration or exposure time of 400 μs, which would theoretically make this device suitable for only lesions in which the vessel diameter is 80 μm or less. Despite this, the laser was found to be safe and effective for most lesions and rapidly became the standard of care for the treatment of port-wine stains. However, only a small percentage of port-wine stains cleared completely (15%–20%), and up to 20 or more treatments were required, in some cases, to achieve this. The majority experienced a significant degree of lightening, and a small percentage of patients, mainly those with more advanced lesions, did not respond at all to treatment (Orten et al., 1996; Kauvar and Geronemus, 1995). This, together with the determination that the appropriate exposure times were between 1 and 10 msec, led to the development of the second-generation flashlamp pumped-dye laser with a much longer exposure time of 1,500 μ (Fig. 10.3). In addition to this, the wavelength of emission has been altered to 590 or 600 nm. The advantages of this are evident. The longer exposure time is an improvement but not ideal. Unfortunately, given the limitations of this form of technology at this point, a longer exposure time is not possible. The longer wavelength

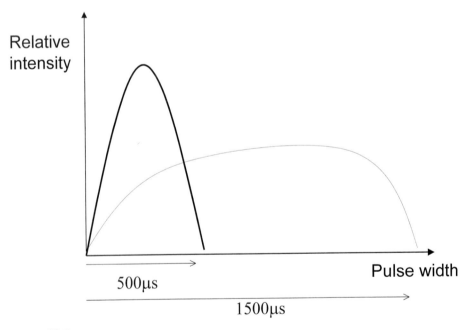

Figure 10.3
Schematic representation of the pulse widths of the first- and second-generation flashlamp pumped-dye lasers.

will result in deeper penetration of the light and hopefully increase the rate of complete response. At this point, there is still insufficient data on the safety and efficacy of these longer pulse widths and wavelengths. However, given that patients in whom a poor response was noted had the bulk of their vessels deeper than 1 mm (Onizuka et al., 1994), it seems likely that these longer wavelengths and longer exposure times will translate into better results.

The copper vapor laser and more recently the copper bromide laser also emit pulsed yellow light at exposure times of between 1 and 100 msec. Unfortunately, the laser emits a wavelength of 578 nm, which corresponds with the $\beta2$ peak in the absorption spectrum of oxyhemoglobin (Fig. 10.1). Light at this wavelength will therefore not penetrate as deep as light at 600 nm. Therefore, while it is capable of producing yellow light at longer exposure times, a longer wavelength would also be desirable.

The only laser capable of producing exposure times in the 1–10-msec range is the KTP laser. This laser, like the copper bromide laser, emits short nanosecond pulses (600–900 nsec) at a rate of 25 kHz (25,000 pulses per second). In addition to this, a shutter mechanism capable of producing millisecond exposure times of between 1 and 100 msec will ensure the correct exposure time. Each millisecond of exposure is thus made up of 25 pulses (Fig. 10.4). Unfortunately, the KTP laser does not

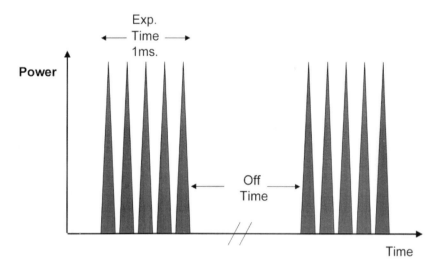

Figure 10.4
Schematic representation of a 1 ms exposure time for a KTP laser. For convenience, only 5 pulses are shown (instead of 25).

emit yellow light. It should not, however, be discounted simply for this reason.

10.1.2. Nonselective Photothermolysis

Although nonselective destruction of vascular tissue has been largely replaced, it still has a role in the management of lesions in which the possibility of selective destruction has not yet been explored. These lesions are made up of large vessels with diameters equal to or greater than 400 μm and include the cobblestones of grade V venular malformations and venous malformation (large vessel disorders). Exposure times of between 100 and 300 msec would be necessary, and because the depth of these vessels is considerably greater than medium and small vessel disorders, a much longer wavelength is preferable. The Nd:YAG laser, which emits a longer wavelength (1,064 nm) and will therefore penetrate to a greater depth (3–8 mm depending on the tissue type), is more suitable for the treatment of this type of lesion (Svaasand et al., 1985). Unfortunately, most commercial Nd:YAG lasers emit continuous wave light, which can be chopped, however, by means of an electromechanical shutter. When used in this way, one loses the advantage of selectivity and nonselectively coagulates the vascular tissue together with some surrounding tissue. The degree of nonselective damage can be limited by selecting an appropriate exposure time. When treating lesions of the tongue, exposure times of 500 msec have been found to be effective while minimizing the degree of scarring. Mucosal lesions should be treated at 200–300-msec exposure times, but treating through skin requires more caution. The potential for scarring is much greater and, therefore, to avoid this, a shorter exposure time of 100 msec is necessary. At the aforementioned exposure times, the laser will ablate submucosal and subcutaneous vascular malformations up to a depth of 0.5–1.0 cm. The reason for this greater depth of effect is the longer exposure time. Once the vascular tissue has been fully saturated with thermal energy, thermal transmission to surrounding tissue will take place because the exposure time is longer than the thermal relaxation time of the vessels.

Treatment in this way can be used as a sole form of treatment or as an adjuvant to surgery. With very superficial venous malformations or advanced (grade V) venular malformations, two or three treatments at least 6 weeks apart will be necessary. With compound lesions, treatment with a laser alone will not be sufficient to destroy the lesion. Subsequent surgical excision will be necessary. Prior treatment with an Nd:YAG laser will, however, result in submucosal fibrosis, which in turn will allow the surgeon to raise a flap using this less vascular plain of dissection (Fig. 11.1). This would have been impossible without prior treatment. In most cases, two or three laser treatments at 6–8-week intervals will be necessary to achieve this degree of fibrosis.

Recent unpublished work has explored the possibility of using a Q-switched Nd:YAG laser (pulse width = 600–900 nm; pulse repetition rate = 25 kHz) chopped to deliver between 20 and 300 msec exposures to treat large vessel lesions. It is hoped that by matching the thermal relaxation time with the appropriate exposure time, selective photothermolysis will be possible.

An innovative way of treating deep venous malformations is by a process known as *interstitial laser photocoagulation* (ILP). Using an optical quartz fiber embedded within the substance of the lesion via a transcutaneous or a transmucosal puncture, light is delivered to the interior of the lesion. If a low intensity of light (1–2 W) is delivered over a long period of time (500–1,000 seconds), the laser light scatters in an isotropic fashion around the emission surface of the fiber and is absorbed and converted to heat, which in turn results in thermal coagulation and necrosis (Mathewson et al., 1987, 1989). Using an Nd:YAG laser and a 600-μm flat cut fiber, a diameter of necrotic tissue of up to 16 mm may result from the point of emission (Mathewson et al., 1987). Earlier work suggested that a high flow rate through large vessels will tend to act as a heat sink, thereby preventing them from reaching a high enough temperature (Mathewson et al., 1987). This can be overcome by allowing the fiber tip to char. Although this will reduce the depth of penetration of light into the surrounding tissue, the light energy will be concentrated into a smaller volume and effect a higher local temperature. Tissue necrosis will thus result from thermal conduction rather than optical penetration of light, which then needs to be converted to heat to destroy the tissue. This mechanism has been found to result in greater thermal damage (Amin et al., 1993a; Mathewson et al., 1987). The extent of necrosis can be controlled by the number of fibers used and their spatial orientation. Experimental work has shown that up to 2 cm of necrosis can be produced using a single fiber and 1,000 J of energy. This will increase to 4 cm when four fibers are placed 1–1.5 cm apart and fired simultaneously (Amin et al., 1993b; Steger et al., 1992). A 1×4 beam splitter is required for this.

Introduction and placement of the fibers under magnetic resonance imaging (MRI) guidance will ensure accurate positioning and holds the potential for real-time monitoring of tissue changes through their changes in MRI enhancement. This would provide exquisite precision in delivering sufficient energy to destroy the lesion with minimal damage to nontargeted tissue. Because one of the determinants of the MRI signal from a given volume of tissue is temperature and because T1-weighted images are particularly sensitive to temperature, it should be possible to monitor temperature changes in real time. Several in vitro and ex vivo studies have correlated MRI signal intensity with temperature and histological changes (Matsumoto et al., 1992; Hushek et al., 1993; Matsumoto et al., 1994). Unfortunately, an in vivo study using rapid image acquisition T1 sequences failed to demonstrate a direct correlation between signal intensity and temperature (Fried et al., 1996). The investigators did, however, note a signal change at the periphery of the surgical site. Although this study does seem to contradict previous reports, the problem may be one of translation of these findings into the in vivo experience. Furthermore, these findings do suggest that it may be possible to monitor the leading edge of a temperature gradient. Additional data will hopefully resolve these issues.

To date, the only published data on ILP using the above-mentioned techniques concerns the treatment of neoplastic disease. Several groups have used a modification of this technique to treat vascular malformations (Gregory, 1991; Alani and Warren, 1992; Apfelberg, 1995). Using a 600-μm optical quartz fiber with 1–2 mm of the distal end of its cladding removed, and the tip sharpened, the fiber is inserted into the substance of the lesion. Several passes are made throughout the lesion using an exposure time of 0.5 second and a power setting of 35 W.

10.1.3. Laser Treatment: A Logical Approach

Lasers can be used to treat superficial vascular lesions and the superficial component of a compound lesion. The vessels may be small, medium, or large, and they can be treated by three types of lasers. None of them is ideal and none of them can be used to treat all types of lesions. The vast majority of work has been done with the flashlamp pumped-dye laser, and this has become the standard of care, especially for the treatment of early grade I and II venular malformations (port-wine stains) that are made up of smaller diameter vessels (Fig. 10.5).

This laser is also appropriate for the treatment of grade III lesions. However, with grade IV lesions, a longer exposure time is necessary because these lesions are made up of vessels with a greater diameter. The two options available are the KTP laser and the copper bromide laser, and both are appropriate. Both of these devices would need to be used with a roboticized scanner to maximize the efficiency and safety of treatment. Large vessel disorders such as venous malformations and the cobblestones of a grade V venular malformation will need to be treated with an Nd:YAG laser.

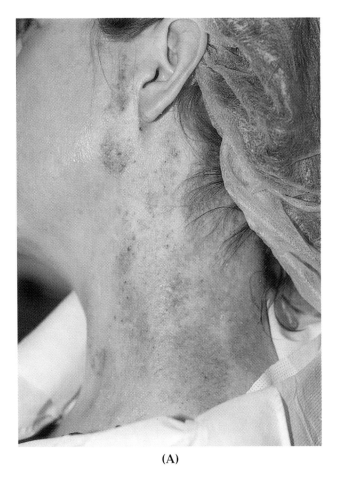

(A)

Figure 10.5
(A) A grade II venular malformation involving the neck of an adult.

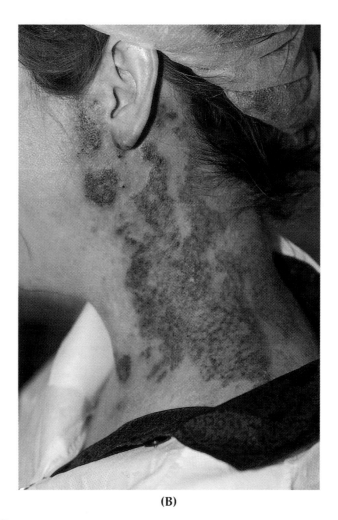

(B)

Figure 10.5 *continued*
(B) The same patient immediately after treatment.

10.2. TECHNIQUES

10.2.1. Small Vessel Disorders

Grades I, II and III venular malformations and midline venular malformations are small vessel disorders. Flashlamp pumped-dye lasers have stood the test of time and are recognized as the laser of choice for the treatment of these disorders. The short pulse (450 μsec) will induce intravascular coagulation with rupture of some of the very small vessels. This coagulum is dark brown but will appear as bluish black when viewed through the collagen of the papillary dermis. The endpoint of treatment is therefore a uniform bluish black discoloration of the entire lesion (Fig. 10.5). In lighter lesions with less oxyhemoglobin this end point is not as obvious immediately after treatment and may take up to an hour to develop. In these circumstances, it may be tempting to increase the energy density in an attempt to achieve this endpoint. This will only increase the attendant risks and should be avoided. Treatment should involve overlapping individual spots by 10%–20% to achieve a uniform end point and hopefully clearance. Contrary to what was previously advocated, this is safe because the distribution of energy across the spot is gaussian (Dinehart et al., 1994). A 7-mm spot size is recommended, and an energy density of between 4.5 and 6.5 J/cm^2 will be sufficient to achieve this end point without risk (Table 10.1). Facial lesions can be safely treated at the higher fluences, but less privileged sites should be treated at lower fluences. When treating a light lesion, erythema, which commences soon after the onset of treatment, will often mask the lesion. The outline of the lesion will need to be marked with a fine-tipped surgical marker prior to the commencement of treatment to prevent this. An alternate strategy would be to carefully note the margins of the lesion prior to treatment and treat the periphery of the lesion first. The center can then be treated in the conventional way.

TABLE **10.1**
The Variation of Fluence With Respect to Spot Size and Wavelength when Treating Venular Malformations

Area	Wavelength (nm)	Spot Size (mm)	Fluence (J/cm^2)
Face (Ped)	585	5	6.0–7.5
		7	5.0–6.5
	595	7	7.0–8.0
Face (Adult)	585	5	6.5–7.5
		7	5.5–6.5
	595	7	8.0–9.0
Neck	585	5	5.0–6.0
		7	4.0–5.0
	595	7	6.0–7.0

The most commonly used technique involves the use of a 5- or a 7-mm hand piece. The fluences listed below are those one should use with a 7-mm handpiece at 585 nm (refer to Table 10.1 for variations).

- Holding the handpiece perpendicular to the skin surface at all times, a series of consecutive pulses are delivered.

- The pulses should overlap by 10%–15% to ensure a uniform distribution of light.

- If the lesion is light, the perimeter of the lesion should be treated first to avoid not being able to discern the edge of the lesion.

- Facial lesions can be treated at up to 6.5 J/cm^2, neck lesions at 4.5–5 J/cm^2, and lesions involving any other anatomical area at 5–5.5 J/cm^2.

- An end point of bluish-black discoloration of the entire treated area should appear after treatment and will deepen in color over the ensuing 24-hour period (Fig. 10.6).

Appropriate eye protection should be worn by both the physician and the patient at all times. Most of the commonly used goggles will not provide adequate protection against direct exposure, so these should not be relied on for patient protection during treatment of the head and neck. The patient's eyes should preferably be taped shut. Should it be necessary to treat the patient's eyelid, an opaque corneal shield should be used. One should always ensure that the shield is large enough to completely cover the cornea and not drift off, leaving it unprotected during treatment.

The treated area should heal within 2–3 weeks. The bluish black discoloration will slowly fade, leaving a reddish hue by day 12, which itself will regress to a pretreatment skin tone by 2–3 weeks post-treatment. Blistering and scabbing are evidence of epidermal destruction, an indication of overtreatment. In this case, the energy density should be reduced by at least 0.5 J/cm^2 during any subsequent treatment. Dressings are only necessary if the treated area is likely to be traumatized during the ensuing 2 weeks. Postoperative management should include application of a thin film of antibiotic ointment, applied as often as is necessary to keep the area moist. An alternative and useful strategy is to apply a refrigerated hydrogel dressing immediately post-treatment. This will relieve the stinging sensation during the first few hours post-treatment. Sun exposure should be avoided, especially during the first 6–8 weeks post-treatment. Routine use of sun protection creams should be encouraged, especially during this period, and the patient should not traumatize the area during the healing phase.

Complications include hyperpigmentation, hypopigmentation, atrophic scarring, and hypertrophic scarring. For a complete discussion of these, refer to Chapter 8.

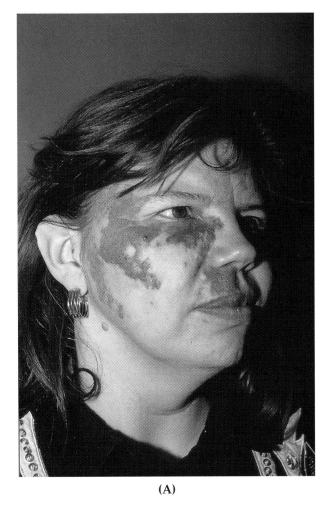

(A)

Figure 10.6
(A) An adult with a grade IV venular malformation in the V2 distribution.

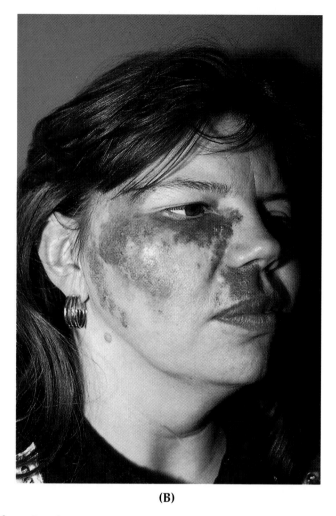

(B)

Figure **10.6** *continued*
(B) The same patient immediately after laser treatment of the lateral portion with a copper vapor laser. Note the end point: a combination of vasoconstriction as well as a blue-black discoloration.

Pulsed-dye lasers are safe and effective in treating skin types 1, 2, 3 and 4 but are contraindicated for skin types 5 and 6 (light and dark black skin). The competitive absorption of this wavelength of light by melanin becomes important and in the presence of a higher concentration of melanin will result in epidermal destruction. The consequences of this are a higher risk of scarring and hypopigmentation.

10.2.2. Medium Vessel Disorders

Medium vessel disorders includes grade IV venular malformations and some early venous malformations. Exposure times of 1–100 msec are required to effectively coagulate these disorders and both the KTP laser and the copper bromide laser are capable of this. Both lasers are nanosecond pulsed lasers with a kHz pulse repetition rate (several thousand cycles/second). With the aid of an electromechanical shutter, one is able to vary the exposure time (on time) between these limits in 1-msec increments with the KTP laser and in 5-msec increments with the copper bromide laser. The technique we describe is a "freehand" technique that has been successfully used by several experienced clinicians. This technique is more difficult than that employed with the flashlamp pumped-dye laser and should not be undertaken by one inexperienced or unschooled in the use of these lasers. One therefore should observe several cases prior to undertaking this form of treatment to minimize the complications.

It is recommended that a test treatment be carried out prior to commencement of the treatment. The purpose of this is to establish the parameters necessary for treatment as well as the safety of further treatment. A representative area of the lesion is chosen, and, following the procedure outlined in the proceeding section, a small area should be treated. The patient should be seen at weekly intervals for 2 weeks and then every other week for at least 6 weeks. By this time the area should have healed completely with no residual swelling, scarring, or residual erythema. Should this not be the case, further treatment should be delayed until complete resolution and a satisfactory explanation as to why healing was delayed. If necessary, a second test patch should be carried out before proceeding with treatment. Scarring or any other major adverse reaction will preclude further treatment.

The treatment technique involves delivering a series of exposures adjacent to each other and looking for an end point of vasoconstriction without blanching the overlying skin. If this is not possible, an end point of bluish black discoloration not dissimilar from that seen with a flashlamp pumped-dye laser is acceptable. The treatment is usually carried out under local anesthesia because the vast majority of patients are adults. Children will obviously need to be treated under general anesthesia.

With the *copper bromide laser,* the maximum power output of the laser is 2 W of light at 578 nm (yellow light) and the spot size is 800 μm (0.8 mm).

- Using the maximum power setting and holding the handpiece perpendicular to the surface of the lesion and at a distance of 1 mm from the surface, a series of exposures are delivered adjacent to each other.
- Starting at an exposure time of 15 msec and increasing it in increments of 5 msec, the appropriate end point should be sought (vasoconstriction of bluish black discoloration without blanching the overlying skin) (Fig. 10.5).
- Once the appropriate exposure time has been found, the entire test area or, if this has already been done, the entire lesion should be treated.
- The speed with which the pulses are delivered can be varied according to how fast the physician wishes to work. An average speed and perhaps a good starting point is five pulses per second.

In most cases, postoperative changes consist of mild skin exfoliation. If the patient should experience blistering or scab formation, too high a fluence was used. Under these circumstances frequent applications of an antibiotic ointment or Vaseline are necessary to promote moist wound healing.

The technique we recommend when using the *KTP laser* is similar. The laser has an aiming beam, and the handpiece is microlensed to provide a focused 1-mm spot at the end of a spacer. The end point is identical, and both the fluence and the exposure time can be adjusted in 1-unit increments to achieve this. As is the case with the copper bromide laser, a test patch should be carried out, and at least 6 weeks should lapse between this and the main treatment. The patient will require local anesthesia (if an adult) or general anesthesia (if a child).

- Keeping the fluence constant between 10 and 15 J/cm^2 and varying the exposure time, the same end point should be sought.
- Once this has been determined, a series of exposures are delivered adjacent to each other until the entire lesion has been treated.
- The handpiece should be kept perpendicular to the surface of the lesion and at the distance of the spacer.

· The speed with which the treatment is carried out will depend on the "off time" or the time between exposures. This can be set at 1–10 pulses per second depending on the speed with which the physician feels comfortable.

As with the copper bromide laser, post-treatment changes consist of mild skin exfoliation. Scabbing or blistering is indicative of too high a fluence and should be treated with frequent topical applications of petrolatum or an antibiotic ointment.

Wherever possible, one should use a computerized robotic scanner to reduce the likelihood of user error and thus the risk of complications. Because both the KTP and the copper bromide laser are user dependent and require considerable experience, computerized scanners will shorten the learning curve, minimize the risks and lead to more uniform results. Computerized scanners are currently available with both the KTP laser and the copper bromide laser.

10.2.3. Large Vessel Disorders

Large vessel disorders include grade V venular malformations (cobblestoned and/or hypertrophied port-wine stains) and most venous malformations. Vessel sizes of between 400 μm and several millimeters require a much longer exposure time. Unfortunately, most of this work is in the realm of nonselective photothermolysis, but recent work with a Q-switched Nd:YAG laser shows promise. The technique we recommend involves the use of a 600-μm flat optical quartz fiber and chopped, continuous wave light.

· The fiber is held perpendicular to the surface being treated.

· The power setting we generally recommend is 30 W.

· An exposure time of between 0.2 and 0.5 second may be selected for the treatment of mucosal surfaces. Tongue lesions usually require longer exposure times (0.5 second because these lesions tend to be deeper and the tongue is less likely to scar).

· Mucosal lesions may be treated at 0.2- or 0.3-second exposure times.

· The vermilion border should be treated at 0.2-second exposure times.

- We prefer a shorter exposure time (0.1 or 0.2 second) for the treatment of subcutaneous lesions.

Individual impactions should be delivered and these should be placed about 2 mm apart. The entire surface should be treated in a grid fashion. If the vascular component is superficial, considerable vasoconstriction will be seen immediately following the delivery of each treatment spot. An end point of pinpoint blanching at each of the treatment spots is desirable (Figs. 10.7 and 10.8).

Because the epicenter of the thermal effect is just below the surface of the mucosa or skin, any more than this is excessive and will result in confluent necrosis. Unfortunately, there is a tendency to overtreat because it may seem natural to rely on a visual end point of confluent blanching. Unlike treatment with other lasers, the effects seen after a few treatments with the Nd:YAG laser are far greater than what is apparent immediately after treatment.

Cutaneous surfaces then should be kept covered with a thin film of petrolatum or an antibiotic ointment. Mucosal surfaces need no special treatment apart from a soft diet.

More than one treatment is usually required. The interval between treatments should not be less than 6 weeks and not greater than 3 months.

Postoperatively, marked swelling will be evident. This will reach maximum proportions within the first 12–18 hours. The perioperative administration of steroids will, however, reduce the degree of swelling. This is especially important when treating lesions involving the oral cavity or airway. The swelling should subside by the second week, and visible improvement should become evident by then. However, because the area heals with fibrosis, it may take several months before the full benefit of the treatment is obvious.

With aggressive treatment, one often finds areas of mucosal separation with blister formation. Re-epithelialization will invariably take place, and, provided the extent of submucosal necrosis is not too extensive, the result will be satisfactory. This degree of aggressive treatment of subcutaneous lesions invariably will result in scar formation and therefore should be avoided.

Postoperative pain is surprisingly minimal and easily controlled with acetaminophen. One should avoid nonsteroidal analgesics because the risk of a secondary hemorrhage between days 6–8 could, theoretically, be greater.

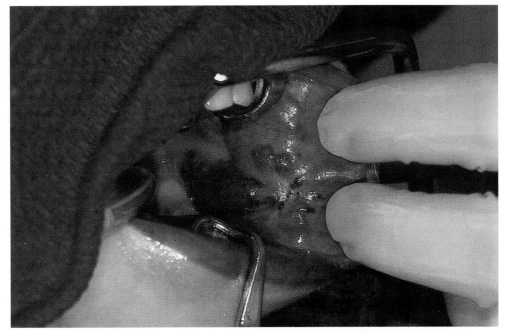

(A)

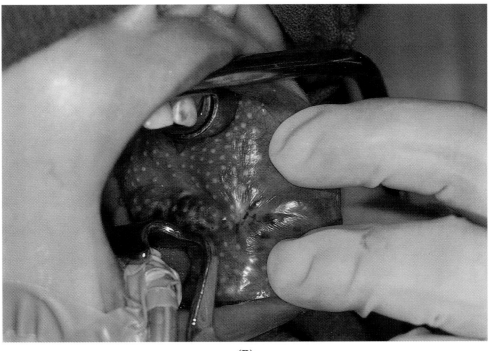

(B)

Figure 10.7
(A) A 3-year-old child with an extensive venous malformation involving her buccal mucosa. **(B)** The same child immediately after treatment of the posterior portion with a Nd:YAG laser. Vasoconstriction is obvious from the color change. Note also each treatment point represented by a circular blanched 1-mm spot.

335

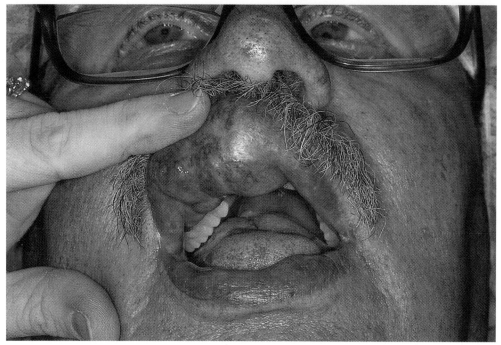

(A)

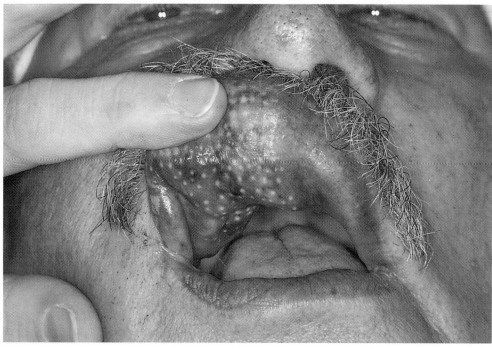

(B)

Figure 10.8
(A) An adult with a venous malformation of his upper lip. (B) The same patient immediately after treatment. Note the vasoconstriction as well as the blanched macule, which represent each area of impaction.

10.3. SCLEROTHERAPY

The use of sclerosing agents for the treatment of certain types of vascular malformations is a viable alternative and indeed a preferred modality in certain circumstances. Sclerosing agents have been used successfully to treat venous malformations and lymphatic malformations, and, within these confines, they can be used to treat both localized and diffuse lesions (de Lorimier, 1995; Riche et al., 1983; Yakes et al., 1989; Dickerhoff and Bode, 1990; Ogita et al., 1991; Gericke et al., 1992; Molitch et al., 1995). When treating localized lesions, the goal is complete obliteration of the vascular channels. However, localized lesions are also amenable to surgical resection, and because it is thought that recannalization is likely (de Lorimier, 1995), perhaps these are the types of lesion that should be excised. When treating diffuse lesions, multiple treatments are usually necessary and may be given over a period of several years. Under these circumstances, our aim is "control" of the lesion. Despite this, in the hands of an experienced interventional radiologist, near-complete ablation of the malformation can be expected. However, the patient should be informed of the probability of a recurrence. Sclerotherapy will therefore at best control the lesion. The only contraindications to sclerotherapy are

- The possibility that one is dealing with a low-grade arteriovenous malformation. An accidental intraarterial injection of sclerosant material may lead to extensive necrosis.
- Direct communication between the venous malformation and the cavernous sinus. To avoid this, it is imperative to administer a test dose of the sclerosing material prior to the final injection.

Several agents have been used as sclerosants: sodium morrhuate 5%; sodium tetradecyl sulfate; ethenolamine oleate; ethyl alcohol; hypertonic saline; prolamine, and amidotrizoic acid, and oleum papaveris; bleomycin; dextrose; tetracycline/doxycycline; OK432 (lyophilized incubation mixture of low virulent Su strain type III, group A Strep. P with Pen. GK). The choice of agent will depend on the experience and preferences of the physician and the type of lesion being treated. Unfortunately, there are no published clinical trials comparing the efficacy and side effects of these agents. Some of the agents do appear to be more effective with certain types of lesion.

10.3.1. Lymphatic Malformations

The cumulative experience with treatment of lymphatic malformations is limited and the results inconsistent. Dextrose appears to be ineffective, whereas excellent results have been reported with OK432 and bleomycin (Okada et al., 1992; Tanigawa et al., 1987; Ogita et al., 1991). Doxycycline also appears to be promising (Gericke, 1989; Molitch, 1995). The mechanism of action common to all of these agents is the diffusion of the agent into the stroma. This results in an irritation that in turn leads to inflammation, sclerosis, and cicatricial contraction of the lesion (Yura et al., 1977; Tanigawa et al., 1987). OK432, on the other hand, appears to diffuse less widely and thereby confine its damage to the cyst wall and lining (Ogita et al., 1991).

The results of treatment with OK432 appear to be consistent, but the numbers small. Between 43% and 55% responded with complete disappearance of their lesion and no evidence of recurrence during a 3-year follow up (Ogita et al., 1991; Yura et al., 1977; Tanigawa et al., 1987; Okada et al., 1992). Multiple treatments were necessary, and it appears that those patients who responded well did so with only one, two, or three treatments. These patients appeared to have macrocystic lesions, which also respond so well to surgery. Patients who required more treatments did so to achieve a measure of control of the lesion and will require periodic treatments to maintain the result. These patients appeared to have extensive microcystic lesions.

The technique used will depend on the agent and physician administering the agent. The most commonly used agents are tetracycline/doxycycline, bleomycin, and OK432. In general,

- An attempt is made to aspirate as much lymph as possible from the lesion. If the mass is composed of multiple noncommunicating cavities, an attempt is made to empty each of these.

- The volume of sclerosant will depend on the agent being used, the nature of the lesion (macrocystic or microcystic), and the weight of the patient. The reader is referred to the referenced articles for this information.

- Injection is usually performed under conscious sedation or with the patient anesthetized because administration of these agents is painful.

- Patients are usually admitted postoperatively to monitor and control complications.

- Adverse effects were common to all agents and included moderate to high fever, local swelling and tenderness, and local pain.

Currently, most authors agree that the primary treatment modality for lymphatic malformations is surgical resection (Hancock et al., 1992; Ogita et al., 1991; Fonkalsrud, 1986). The role of sclerotherapy therefore appears to be relegated to the treatment of diffuse or persistent areas or lesions in which surgical resection is likely to disfigure the patient. This however may change. An increasing body of favorable data with OK432 may well mitigate in favor of sclerotherapy. Sclerotherapy has also been advocated as a primary modality for localized macrocystic lesions, the type that also respond well to surgical excision. The major advantage is the absence of a scar. Several disadvantages do, however, exist. These include the fact that multiple treatments are necessary, patients may need to be hospitalized post-treatment (this depends on the agent being used), and significant side effects may occur. Over the course of the next few years, this issue will undoubtedly be resolved.

10.3.3. Venous Malformations

The cumulative experience is, again, limited, and the most commonly used agents include sodium tetradecyl, absolute alcohol, sodium morrhuate, ethanolamine, and Ethibloc (Yakes et al., 1989; de Lorimier, 1995; Riche et al., 1983; Kauffmann et al., 1981). Because Ethibloc has a very slow solidification time (10–15 minutes), venous tourniquets have been used to prevent distal migration of the substance during the treatment of venous malformation of the extremities. Flow arrest in the head and neck using this technique is impossible. Distal migration of the agent and unwanted secondary effects therefore all mitigate against this agent being used in the head and neck.

One of the main considerations with regard to head and neck lesions is the fact that veins of the head and neck lack valves, and much of the middle third of the face communicates with the cavernous sinus by way of the superior and the inferior ophthalmic veins. The possibility of accidental thrombosis of the cavernous sinus therefore should always be kept in mind when considering sclerotherapy. For this reason, we do not advocate sclerotherapy for lesions of the upper or middle third of the face. On the other hand, most of the venous flow from the lower third of the face is directed inferiorly toward the jugular system. Sclerotherapy of lesions of the lower third of the face and the neck is therefore safer.

The technique we describe has been adapted from that of de Lorimier (1995) and can be used with any of the commonly used agents with the exception of Ethibloc.

- All treatments should be administered under general anesthesia, because the administration of a sclerosing agent is painful.
- A 25-gauge butterfly needle (25-gauge, 0.75-inch length; Abbott Laboratory, North Chicago) should be introduced into one of the venous spaces and taped in position. The fine plastic tubing attached to the needle should then be connected to a three-way stopcock.
- The location of the puncture site and the extent of filling of the malformation is then carefully checked by the injection of contrast medium under fluoroscopic observation. This is critical to ensure that the sclerosing agent is given intravenously and not into the surrounding subcutaneous tissues. It is also important to establish that the sclerosing agent fills the venous malformation and does not pass directly into a large draining venous channel. Fluoroscopic analysis will also ensure that there is no flow via the superior or inferior ophthalmic vein into the cavernous sinus. By simply injecting a test dose of contrast medium, the flow pattern can be established.
- The sclerosing agent is then injected until the venous channel becomes engorged and thrombosed and until blood no longer returns with aspiration. If the lesion is superficial, the injection will be stopped when it appears that the sclerosing agent has entered normal cutaneous vessels. This minimizes the risks of skin necrosis.
- The volume of sclerosing agent used depends on the agent and the extent of the venous malformation. As a rule, not more than 1 ml/kg body weight should be used.
- Since hemoglobinuria is frequently noted, an intravenous infusion of lactated Ringer's solution should be administered at twice the normal maintenance rate to avoid renal tubule damage.

The number of treatments required per lesion will depend on the extent of the lesion. Small localized lesions will only require a single treatment; however, in most instances, repeat treatments will be necessary. These should be administered at 4–6-week intervals to allow the induration and inflammation to subside.

At this point, the precise role of sclerosing agents in the treatment of malformations has not been established. When used as a adjunct to surgery, sclerosing agents are extremely useful. When used as the sole modality of treatment, recanalization of the vessels over a period of time with recurrence of the lesion is common (de Lorimier, 1995). It is therefore thought that in most diffuse lesions sclerosing agents will at best control the lesion.

Complications will depend on the degree of aggression of the physician. When injecting small volumes of sclerosing agents, the rates of complications will be lower, but multiple treatments will be necessary because efficacy is reduced. Unfortunately, there are no clinical comparative trials so the choice of agent will be that of the physician in charge of the case. The following complications have been recorded:

- Allergic reactions: Anything from mild rash to urticaria to anaphylaxis should be anticipated when fatty acid–based solutions or tetradecyl sulfate is used.

- Cerebral intoxication: Absolute alcohol has not resulted in allergic reactions, but the volume of the injection must be limited to avoid cerebral intoxication. The maximum volume of absolute alcohol for a 70-kg individual should be 1.2 ml. Absolute alcohol should not be used to treat children. If large volumes are needed, absolute alcohol should not be the choice of agent.

- Skin necrosis: This is a frequent complication. In one series it occurred in 10% of patients (de Lorimier, 1995). This presumably results from extravasation or reflux of the sclerosing agent into the subcutaneous tissue.

- Neuropraxia: Extravascular injection of the sclerosant or treatment of a venous anomaly surrounding a motor or sensory nerve will result in neuropraxia of that nerve.

10.4. SURGICAL RESECTION

Surgical resection can, at times, be extremely difficult. The main challenges are hemostasis, complete removal of the lesion, preservation of important structures, and establishment of normal facial appearance. In general, given the progressive nature of vascular malformations, it is far easier to operate at an early stage than at a later stage because both the degree of ectasia and the flow rate are considerably less. By the same token, one should also take into account the hazards of operating on a small child. The total circulating blood volume of an infant is considerably less than that of an adult, and, given the risks of intraoperative hemorrhage, a compromise should be sought.

10.4.1. Hemostasis

The surgeon should be familiar with the use of the contact laser or thermoscalpels prior to undertaking resection of a vascular malformation. Both have been found to be extremely useful and are considered essential to the successful removal of vascular malformation (refer to Chapter 8 for a more complete discussion on these devices).

Of all the vascular lesions, the maintenance of hemostasis is most difficult with venous malformations because these vessels are extremely thin walled and very often severely ectatic. Accidental violation of the wall of one of these vessels can result in profuse hemorrhage, which is difficult to control. Conventional electrocautery will usually not succeed because there is insufficient muscularis media in the wall of the vessel and connective tissue within the stroma of the malformation to thermally contract and stop the bleeding. The large diameter of the vessels also precludes successful hemostasis with electrocautery. One may therefore have to resort to the use of hemoclips or hemostatic sutures. An accidental hemorrhage close to a structure such as one of the branches of the facial nerve can be particularly bothersome. The branch of the facial nerve will need to be dissected free and then mobilized away from the area of hemorrhage before an attempt to control the hemorrhage can be made. In this way, accidental injury to the nerve can be prevented.

Preoperative embolization will be helpful in reducing blood loss and should be undertaken 24 hours prior to resection of an arteriovenous malformation. With regard to venous malformations, preoperative sclerotherapy may have a similar effect. The sclerosing agent will thrombose the vessels, and, if the entire lesion has been sclerosed, the margins of the lesion will also be more readily defined. Unfortunately, sclerotherapy is still unsafe in lesions of the middle and the upper third of the face because facial veins have no valves and the venous drainage may well pass through the cavernous sinus. Sclerotherapy in these circumstances should be undertaken with extreme caution and under direct screening.

10.4.2. Extent of Resection

Unlike involuting hemangiomas, complete resection of a vascular malformation is essential to prevent recurrence. It is therefore necessary to delineate the full extent of the lesion prior to surgery. An MRI will be sufficient in most cases, but, with respect to arteriovenous malformations, it will also be necessary to evaluate the findings of digital subtraction angiography (DSA). Furthermore, it is important for the surgeon to be present during the angiogram so that he may appreciate the flow dynamics of the lesion as well as the extent of the nidus because both pieces of information are helpful during surgery. Combining the findings of an MRI with DSA will help the surgeon to fully appreciate the full extent of the nidus and thereby reduce the likelihood of leaving a potential source of recurrence (Fig. 10.9).

Although complete removal should be attempted whenever possible, when dealing with a diffuse lymphatic malformation or a very extensive venous malformation, this may not be possible without compromising the function and/or the facial features of the patient. Under these circumstances, serious consideration should be given to a partial resection. It is always possible to re-visit the site surgically or with interventional radiology if this becomes necessary at a later stage. In the presence of extensive venous malformations it may also be necessary to stage the resection because complete resection may take as long as 24 hours to accomplish. This is especially true of extensive cervicofacial lesions involving the parotid, buccal fat space, parapharyngeal space, and the anterior triangle of the neck. Multiple transfusions of blood products may be necessary, and eventually these will have an adverse effect on coagulation, which will further compound the difficulty of the resection.

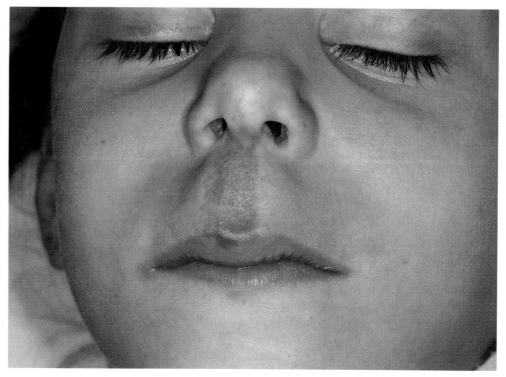

Figure 10.9
An arteriovenous malformation in the midline of the upper lip. Incomplete excision resulted in an early recurrence of the lesion. A subsequent MRI and DSA demonstrated nidus in the floor of the nose as well as involvement of the inferior turbinates. Complete resection eventually took place and this child has no recurrence.

10.4.3. Preservation of Important Structures

The preservation of important structures such as the facial and hypoglossal nerves, the muscles of facial expression, and any of a number of other structures should be a primary concern during surgery. It is possible to remove extensive venous malformations or lymphatic malformations while preserving these structures (Fig. 10.10). Careful dissection, meticulous hemostasis, and the aid of magnification are helpful.

A common site for both lymphatic malformations and venous malformations is the buccal fat space, and removal of the lesion from this location exemplifies these principles. The buccal fat space is limited by the buccal mucosa medially, the skin and subcutaneous fat laterally, the zygomatic arch above, and the ramus of the mandible inferiorly. The

Figure **10.10**
A view of the buccal fat space showing the lower division facial nerve dissected out and mobilized prior to removal of the remaining venous malformation.

masseter muscle forms the posterior margin and the upper and lower lips the anterior margin. The space contains buccal fat and the zygomatic and buccal branches of the facial nerve. Lesions in this location are extremely difficult because resection will place these facial nerve branches at risk. Careful dissection of the branches of the facial nerve through the fat space, followed by mobilization of these branches, is initially undertaken. This may be extremely difficult especially in the presence of a venous malformation. Once this has been accomplished, the vascular malformation that has usually replaced the fat space can then be removed. This space is continuous with the superficial temporal space and passes deep to the zygoma between the superficial temporal fascia and the temporalis muscle to terminate behind the lateral wall of the orbit.

10.4.4. Establishment and/or Preservation of Normal Facial Features

Unfortunately none of the vascular malformations respect tissue planes. Instead, they tend to infiltrate muscle frequently, making it difficult to extirpate the lesion without sacrificing at least some muscle. With regard to both lymphatic and venous malformations, the grade of the malformation will dictate the extent of resection. In the presence of a high-grade lesion, early recurrence can be expected, and serious consideration should be given to complete resection even if this is likely to result in some disfigurement. On the other hand, in the presence of a low-grade lesion, the extent of resection can be less complete. It is therefore imperative that the surgeon evaluate all aspects of the case prior to making a decision. This may require consultation with the child's parents, a social worker, and a psychologist. A full understanding of the natural history and all other factors associated with vascular lesions is therefore imperative and will help the surgeon reach a decision.

10.5. EMBOLIZATION

Embolization has been found to be particularly useful in the management of arteriovenous malformations. Unfortunately, despite the advances that have taken place during the last few years, embolization as a sole means of treatment must be considered a temporary measure because the blood supply to a defined anatomical area is not static and the development of a colateral blood supply is inevitable. Embolization can therefore be used as a preoperative strategy to reduce intraoperative blood loss or as a sole form of treatment for inoperable lesions with the understanding that this is at best palliative. Palliative embolization can also be delivered in stages without greatly increasing the morbidity.

The aim of embolization should be to occlude the nidus of the malformation and not the feeding vessels. Embolization should therefore proceed from distal to proximal to ensure this. The invariable establishment of a collateral circulation will depend on previous manipulations as well as the hemodynamic requirements of the vascular lesion. An appreciation of the usual blood supply to a lesion is therefore important.

A transfemoral approach is used for most facial lesions, and the particles should be delivered via a 4 or 5 French gauge catheter with its tip positioned as close to the lesion as possible. It is important that the catheter should not occlude the vessel; this will prevent the required distal inflow of the particles. The choice of particle will depend on whether the patient is being treated palliatively or as a presurgical strategy. Polyvinyl alcohol foam has a permanent occlusive capacity, whereas Gelfoam only provides temporary occlusion. The use of absorbable gelatin sponge particles together with polyvinyl alcohol foam reduces the coefficient of friction, and this facilitates distal delivery of a greater number of the polyvinyl alcohol foam particles. For effective embolization, it is critical to obliterate the vascular spaces within the substance of the nidus. This is then followed by more distal to proximal occlusion of the arterial blood supply. The delivery of embolic particles should be monitored fluoroscopically, and the gradual build up of peripheral resistance must be recognized.

Preoperative embolization is usually undertaken with Gelfoam Powder as the embolic agent. Gelfoam is chosen to minimize the potential risks associated with accidental migration of the embolic material embolization. The particle size is approximately 40 μm, and the Gelfoam is suspended in a dilute contrast material at the time of embolization. The Gelfoam slurry is drawn into 1-cc syringes for injection. Once again, real-time monitoring in the form of DSA is essential for both safety and efficacy and should be undertaken in every case.

Potential complications include the backflow of embolic material into the intracranial circulation, which can lead to a cerebrovascular accident. Aggressive embolization may result in necrosis of normal tissue. An example of this is necrosis of facial skin, which can occur with inadvertent occlusion of the facial artery. Gelfoam particles minimize the risk of permanent neurological damage because they are reabsorbable. The likelihood of complications is thus significantly reduced after embolization.

REFERENCES

Alani, H., and Warren, R.: Percutaneous photocoagulation of deep vascular lesions using a fiberoptic wand. Ann. Plast. Surg. 29:143–148, 1992.

Amin, Z., Bown, S., and Lees, W.: Liver tumor ablation by interstitial laser photocoagulation: Review of experimental and clinical studies. Semin. Intervent. Radiol. 10:88–100, 1993a.

Amin, Z., Donald, J., Master, A., et al.: Hepatic metastases: Interstitial laser photocoagulation with real-time US monitoring and dynamic CT evaluation of treatment. Radiology 187:339–347, 1993b.

Anderson, R., and Parrish, J.: Selective photothermolysis: Precise microsurgery by selective absorption of pulsed radiation. Science 220:521–527, 1983.

Apfelberg, D.: Intralesional laser photocoagulation-steroids as an adjunct to surgery for massive hemangiomas and vascular malformations. Ann. Plast. Surg. 35:144, 1995.

de Lorimier, A.: Sclerotherapy of venous malformations. J. Pediatr. Surg. 30:188–194, 1995.

Dickerhoff, R., and Bode, U.: Cyclophosphamide therapy in nonresectable cystic hygroma. Lancet 335:1474–1475, 1990.

Dierickx, C., Casparian, M., Venugopalan, V., Farinelli, W., and Anderson, R.: Thermal relaxation of port-wine vessels probed in vivo: The need for 1–10-millisecond laser pulse treatment. J. Invest. Dermatol. 105:709, 1995.

Dinehart, S., Flock, S., and Waner, M.: Beam profile of the flashlamp pumped pulsed dye laser: Support for overlap of exposure spots. Lasers Surg. Med. 15:277–280, 1994.

Fonkalsrud, E.: Disorders of the lymphatic system. In Welch, K.J., Randolph, J. G., Ravitch, M.M., et al. (eds.): *Pediatric Surgery.* 4th Ed. Chicago: Year Book Medical, 1986, pp. 1506–1507.

Fried, M., Morrison, P., Hushek, S., Kernahan, G., Jolesz, F.: Dynamic T1-weighted magnetic resonance imaging of interstitial laser photocoagulation in the liver: Observations on in vivo temperature sensitivity. Lasers Surg. Med. 18:410–419, 1996.

Garden, J.M., Polla, L.L., and Tan, O.T.: The treatment of port-wine stains by the pulsed dye laser: Analysis of pulse duration and long-term therapy. Arch. Dermatol. 124:889–896, 1988.

Gericke, K.: Doxycycline as a sclerosing agent. Ann. Pharmacother. 26:648–649, 1992.

Greenwald, J., Rosen, S., Anderson, R.R., Harrist, T., MacFarland, F., Noe, J., and Parrish, J. A.: Comparative histological studies of the chunaville dye (577 nm) laser and argon laser: The specific vascular effects of the dye laser. J. Invest. Dermatol. 77:305, 1981.

Gregory, R.: Treatment of cavernous hemangiomas with intralesional ablation with YAG laser. Lasers Surg. Med. 11(3):66, 1991.

Hancock, B., St. Vil, D., Luks, F., Dilorenzo, M., and Blanchard, H.: Complications of lymphangiomas in children. J. Pediatr. Surg. 27:220–226, 1992.

Hushek, S.G., Morrison, P.R., Kernahan, G.E., Fried, M.P., and Jolesz, F.A.: Thermal contours from magnetic resonance images of laser irradiated gels. Proc. Biomed. Optics. Eur. 2082:52–59, 1993.

Kauffmann, G., Rassweiler, J., Richter, G., et al.: Capillary embolization with ethibloc: New embolization concept tested in dog's kidneys. Am. J. Radiol. 137:1163, 1981.

Kauvar, A., and Geronemus, R.: Repetitive pulsed dye laser treatments improve persistent port-wine stains. Dermatol. Surg. 21:515–521, 1995.

Mathewson, K., Barr, H., Traulau, C., and Bown, S.: Laser photocoagulation: Studies in a transplantable fibrosarcoma. Br. J. Surg. 76:378, 1989.

Mathewson, K., Coleridge-Smith, P., O'Sullivan, J., Northfield, T., and Bown, S.: Biological effects of intrahepatic neodymium:yttrium-aluminum-garnet laser photocoagulation in rats. Gastroenterology 93:550–557, 1987.

Matsumoto, R., Mulkern, R.V., Hushek, S.G., and Jolesz, F.A.: Tissue temperature monitoring for thermal interventional therapy: Comparison of T1 weighted. MR sequences. J. Magn. Reson. Imag. 4:65–70, 1994.

Matsumoto, R., Oshio, K., Jolesz, F.A.: Monitoring of laser and freezing induced ablation in the liver with T1-weighted MR imaging. J. Magn. Reson. Imag. 2:555–562, 1992.

Molitch, H., Unger, E., Witte, C., and vanSonnenberg, E.: Percutaneous sclerotherapy of lymphangiomas. Radiology 1994:343–347, 1995.

Morelli, J., Tan, O., Garden, J., et al.: Tunable dye laser (577 nm) treatment of port wine stains. Lasers Surg. Med. 6:94–99, 1986.

Ogita, S., Tsuto, T., Deguchi, E., Tokiwa, K., Nagashima, N., and Iwai, N.: OK-432 therapy for unresectable lymphangiomas in children. J. Pediatr. Surg. 26:263–270, 1991.

Okada, A., Kubota, A., Fukuzawa, M., Imura, K., and Kamata, S.: Injection of bleomycin as a primary therapy of cystic lymphangioma. J. Pediatr. Surg. 27:440–443, 1992.

Onizuka, K., Tsuneda, K., Shibata, Y., Ito, M., and Sekine, I.: Efficacy of flashlamp pumped dye laser therapy for port wine stains.: clinical assessment and histopathological characteristics. Br. J. Plast. Surg. pp. 271–279, 1995.

Riche, M., Hadjean, E., Huy, T., and Merland, J.: The treatment of capillary–venous malformation using a new fibrosing agent. Plast. Reconstr. Surg. 171:607, 1983.

Rosen, S.: Vascular supply of normal skin and the comparative histologic effects of the tunable dye (at 577 nm) laser and argon laser on normal skin. In K.A. Arndt, J.M. Noe, and S. Rosen (eds.): *Cutaneous Laser Therapy: Principles and Methods.* New York: Wiley, 1983, pp. 53–63.

Steger, A., Lees, W., Shorvon, P., Walmsley, K., and Bown, S.: Multiple-fibre low power interstitial laser hyperthermia: Studies in the normal liver. Br. J. Surg. 79:139–145, 1992.

Svaasand, L., Boerslid, T., and Overaasen, M.: Thermal and optical properties of living tissue: Application to laser-induced hyperthermia. Lasers Surg. Med. 5:589–602, 1985.

Tan, O., Carney, M., Margolis, R., et al.: Histologic response of port-wine stains treated by argon, carbon dioxide and tunable dye lasers. Arch. Dermatol. 122:1016–1022, 1986.

Tan, O., Sherwood, K., and Gilchrest, B.: Treatment of children with port-wine stains using the flashlamp-pumped tunable dye laser. N. Engl. J. Med. 320:416, 1989.

Tanigawa, N., Shimonatsuya, T., Takahashi, K., et al.: Treatment of cystic hygroma and lymphangioma with the use of bleomycin fat embolization. Cancer 60:741–749, 1987.

Yakes, W., Haas, D., Parker, S., et al.: Symptomatic vascular malformations. Ethanol embolotherapy. Radiology 170:1059–1066, 1989.

Yura, J., Hashimoto, T., Tsuruga, N., et al: Bleomycin treatment for cystic hygroma in children. Arch. Jpn. Chir. 46:607–614, 1977.

The Treatment of Vascular Malformations

**MILTON WANER, M.D., F.C.S. (S.A.) AND
JAMES Y. SUEN, M.D., F.A.C.S.**

With regard to vascular malformations, the choice of treatment modality will depend on several factors, which include its vascular content, anatomical location, and depth. Treatment options include laser therapy, sclerotherapy, embolization, and surgical resection. These modalities are not mutually exclusive. More than one modility may be used to treat the same lesion. As a rule, venular and superficial venous malformations can be treated with lasers, whereas arteriovenous malformations and lymphatic malformations are usually surgically excised. Compound venous malformations are usually first treated with a laser and then excised. The rational for prior laser treatment is to create a relatively avascular fibrous plain through which the skin or mucosal flap can be elevated (Fig. 11.1). Extensive arteriovenous malformations

Hemangiomas and Vascular Malformations of the Head and Neck, Edited by Waner, M.D. and Suen, M.D.
ISBN 0471-17597-8 © 1999 Wiley-Liss, Inc.

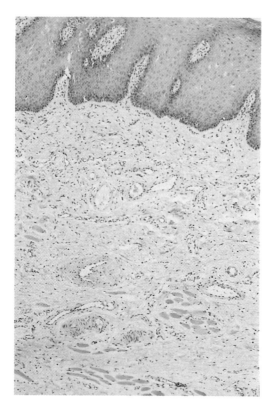

Figure 11.1
A hematoxylin and eosin (H&E)-stained section of a venous malformation after two previous treatments with an Nd:YAG laser. At least 3–4 mm of fibrosis will allow a skin flap to be elevated through a considerably less vascular plain.

are usually embolized preoperatively and then excised because prior embolization will reduce the risk of intraoperative blood loss. The role of sclerotherapy is not yet fully appreciated. At this point, sclerotherapy may be used to palliate inoperative venous malformations and lymphatic malformations or as an adjunct to conventional treatments. With advances in interventional techniques, we may see a shift away from surgical management. Throughout the decision process, one should always bear in mind the natural history of the lesion and choose the most appropriate time for treatment.

11.1. MIDLINE VENULAR MALFORMATIONS

Most midline venular malformations disappear spontaneously by age 6 years (Oster and Nielson, 1970). Furthermore, because these lesions are almost always very light, treatment is usually delayed. Should it become apparent that spontaneous disappearance is unlikely, or should a decision be made to proceed with treatment for whatever other reason, the most appropriate treatment is laser photocoagulation with a flash-lamp pumped-dye laser. These lesions usually respond well and can be expected to resolve completely with one or two treatments (Orten et al., 1996). In general, a low energy density, 6–6.5 J/cm^2, using a 5-mm spot size, will usually suffice. Spots should be overlapped by 10%–20%. The entire lesion should be photocoagulated to an end point of uniform bluish gray discoloration. If a second treatment is necessary, a 3–6-month interval should elapse between treatments.

At these low fluences, few complications are encountered. However, possible complications include postinflammatory hyperpigmentation, hypopigmentation, atrophic scarring, and hypertrophic scarring. Of these, temporary postinflammatory hyperpigmentation is probably the most common. Its precise incidence is not known.

11.2. VENULAR MALFORMATIONS

Unlike *midline* venular malformations, venular malformations never fade; they darken and thicken with age. The mechanism for this is believed to be progressive venular ectasia due to a relative or a complete absence of sympathetic and parasympathetic nerve supply to the papillary venous plexus (Smoller et al., 1986; Rydh et al., 1991; Waner and Suen, 1995). This will result in a loss of venomotor tone, which in turn leads to progressive dilation. The rate of dilation is probably related to the degree of absence of autonomic nerves. A complete absence will result in a more rapid dilation, whereas a relative deficiency will cause a slower dilation. Other factors such as hormonal modulation and trauma will also have an impact.

The choice of treatment depends on the degree of ectasia (vessel diameter; see Chapter 10.) and not on the color of the lesion because the color of a lesion is determined by the degree of oxygenation of the hemoglobin. The degree of oxygenation depends on the degree of perfusion of the capillary bed, which in turn depends on a number of factors such as ambient temperature and the level of circulating catecolamines and local metabolites. The color of a lesion will thus vary throughout the day and is therefore not an appropriate characteristic with which to classify venular malformations. Waner (1989) thus classified venular malformations in accordance with their degree of vascular ectasia. This classification recognized four grades of ectasia, graded I through IV. Grade I represented the smallest vessels and grade IV the largest (see Chapter 3). When using this classification, one should always bear in mind that there is a progression between the grades and that the divisions between these grades are, to a large extent, arbitrary. The main purpose of this classification is to assign a grade for the purposes of communication and determination of treatment modality. Although the color of a lesion is not used as a means for classification, the color will be mentioned as a clinical correlate.

Grade I lesions are the earliest lesions and thus have the smallest diameter vessels (50–80 μm in diameter). Using 6× magnification and oblique lighting, individual vessels can just be discerned. Clinically, these lesions are light or dark pink macules (Fig. 11.2).

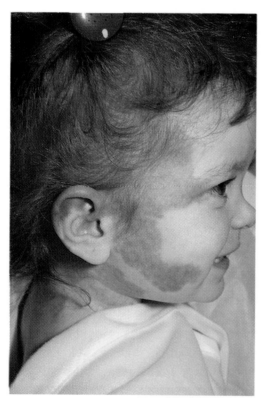

Figure 11.2
A child with a confluent grade I venular malformation.

Grade II lesions are more advanced (vessel diameter is 80–120 μm). Individual vessels are clearly visible to the naked eye, especially in a less dense area. The clinical correlate is a light red macule (Fig. 11.3).

Grade III lesions are more ectatic (120–150 μm). Large end-on vessels are visible and impart a reddish color to the lesion. These lesions are a deeper color and still macular (Fig. 11.4).

Grade IV lesions are the most advanced (>150 μm). By this stage, the space between the vessels has been replaced by the dilated vessels. Individual vessels may still be visible on the edges of the lesion or in a less dense lesion, but by and large individual vessels are no longer visible. The lesion is usually thick, purple, and palpable. Eventually, dilated vessels will coalesce to form nodules, otherwise known as *cobblestones* (Fig. 11.5).

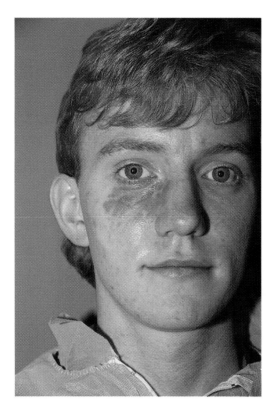

Figure **11.3**
A young adult with a geographic grade II venular malformation.

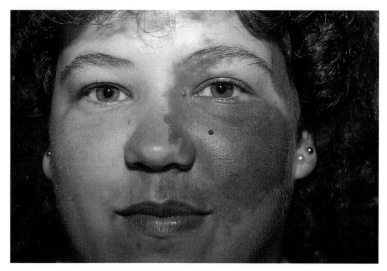

Figure **11.4**
A young adult with a confluent grade III venular malformation.

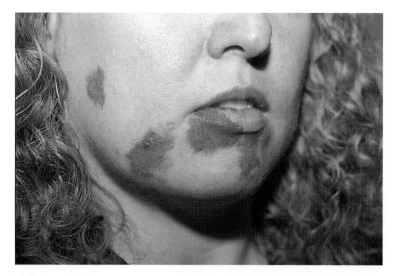

Figure **11.5**
An adult with a grade IV venular malformation. Early cobblestone formation is evident.

Grade I and II lesions are in essence small vessel disorders and are best treated with a flashlamp pumped-dye laser. Grade IV lesions are made up of medium-sized vessels and are best treated with a copper bromide laser or a KTP laser, whereas grade III are made up of vessels that are intermediate in size and can thus be treated with either a flashlamp pumped-dye laser or a copper bromide laser. The only contraindication to treatment is skin type V and VI. The higher concentration of melanin competes for yellow light and results in epidermal destruction. The risk of scarring and hypopigmentation is thus much higher. Because Grades I and II lesions represent the opposite end of the spectrum from grade IV lesions, these two groups are discussed separately.

11.2.1. Grades I and II Lesions

These lesions are made up of small vessels and are best treated with a flashlamp pumped-dye laser. For a comprehensive description of the procedure, refer to Chapter 10. We recommend the liberal use of general anesthesia for children because the procedure is painful and thus very traumatic for a young child. Occasionally one finds a child who will tolerate the procedure without anesthesia, but this is more the exception than the rule.

11.2.2. Grade IV Lesions

Grade IV lesions are made up of medium-sized vessels and are best treated with a copper bromide laser or a KTP laser. The techniques used with these devices are discussed in Chapter 10. Again, the use of general anesthesia is recommended for young children and local anesthesia for adults.

11.2.3. Grade III Lesions

Because both flashlamp pumped-dye lasers and copper vapor lasers/KTP lasers are effective in the treatment of grade III lesions, the laser of choice will depend on availability, convenience, and personal preference. Flashlamp pumped-dye lasers are easier to use, less user-dependent, and faster and are thus the laser of choice for grade III lesions in the authors' hands. Copper vapor lasers and KTP lasers can also be used successfully but are less efficient. This does not preclude their use for grade III lesions. They are currently being used successfully and efficiently, and, with the advent of robotized scanners, their reliability and ease of use have improved. The method of use advocated for all three lasers is described in Chapter 10.

11.2.4. Complications

Adverse events may result from pigmentation changes and/or scarring. Most of the published data concern the complications resulting from treatment with the flashlamp pumped-dye laser (Achauer et al., 1993). Of all the complications, hyperpigmentation is the most frequent and is seen in 10%–20% of patients in the interval between the second and eighth weeks post-treatment. Left untreated, it will resolve spontaneously and may take from 6 weeks to 6 months to do so. The etiology of hyperpigmentation is postinflammatory and it is therefore self-limiting. Postinflammatory hyperpigmentation is more common in olive-complected individuals and after sun exposure. In the face of a high probability, avoidance of sun exposure and prophylaxis with topical 4% hydroquinone is advisable. Hydroquinone can also be used to treat postinflammatory hyperpigmentation and should be used twice daily for up to 6 weeks. Hypopigmentation is another potential complication that is probably under-reported and is only mentioned as an extremely rare event (Geronemus, 1993; Reyes and Geronemus, 1990; Tan et al., 1989). Because it appears as a late complication (6–9 months post-treatment), its true incidence is undoubtedly higher (Fig. 11.6). Furthermore, given the large number of treatments proposed by at least one group, this adverse effect will probably be seen with greater frequency (Kauvar and Geronemus, 1995). Both atrophic and hypertrophic scarring are rare (Fig. 11.7). Their incidence is given as less than 1%, and they are probably more common in unpriviliged sites such as the neck and chest (Reyes and Geronemus, 1990).

Unfortunately, there are very little published data on the complications of treatment with a copper bromide laser or a KTP laser. Because these lasers are used at longer exposure times and higher fluences and because they appear to be more user-dependent and less user-friendly, their incidence of complications is likely to be higher. However, because the use of robotic scanners has replaced the freehand techniques

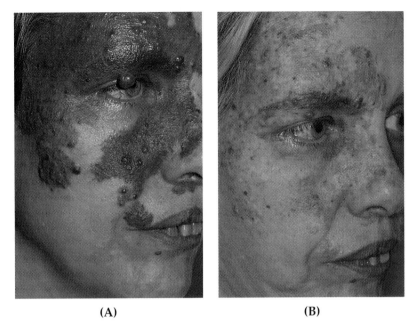

(A)   (B)

Figure 11.6
(A, B) An adult with a cobblestone grade IV venular malformation showing evidence of hypopigmentation after treatment with a copper vapor laser and a flash-lamp pumped-dye laser. The hypopigmentation only became evident 6–9 months after the last treatment.

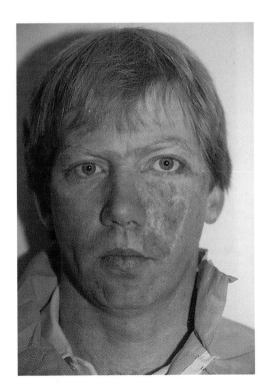

Figure 11.7
A young adult with atrophic scarring and hypopigmentation after aggressive treatment with a copper vapor laser.

used by older physicians, the likelihood of these complications should only be marginally higher.

11.2.5. Response

It is most unfortunate that despite the large number of patients who have been treated and the fact that these lasers have been around for more than 10 years, reports on the response rate remain controversial. Tan et al. (1989) reported a 100% response rate in a series of 80 patients. This publication drew an immediate response. Brauner (1989) believed this to be an inaccurate reflection, as did Garden (1989). Orten et al. (1996) reported only a 15% complete response rate, while the majority of patients (65%) experienced considerable improvement (Figs. 11.8–11.11). Nineteen percent responded with little or no improvement.

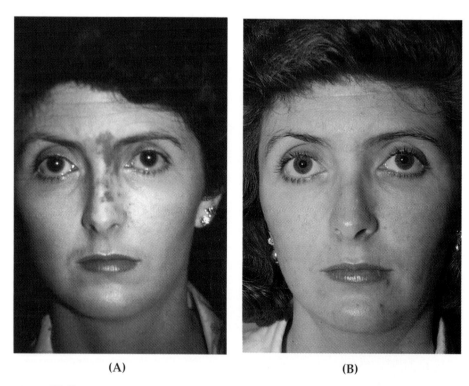

(A)   (B)

Figure **11.8**
(A, B) An adult with a grade III venular malformation that had responded completely to treatment with a pulsed-dye laser. The patient was treated four times with an energy density of 6.5–7.5 J/cm^2 (5-mm spot size).

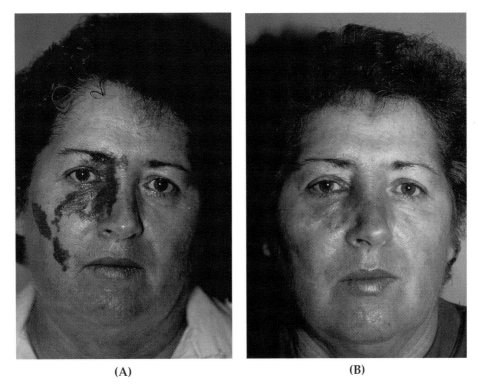

(A)   (B)

Figure **11.9**
(**A, B**) An adult with a grade IV venular malformation that responded well to treatment with a copper vapor laser initially and then with a pulsed-dye laser. The energy
densities used were well within the normal ranges. Note that the lesion failed to clear
completely and instead left a faint pink residuum despite aggressive treatment.

An average of five to six treatments were necessary to achieve maximal
clearance, and children did not respond better than adults. Some of
these patients required up to 20 treatments.

An important question concerns the reason why so many patients
remained with a residuum. This may be due to the fact that some of the
vessels were too deep and thus beyond the depth of penetration of yellow light. Alternatively, some of the vessels may remain unaffected for
an unknown reason. Interestingly, Onizuka et al. have shown a direct
correlation between the depth of the lesion and the likelihood of its
clearance (Onizuka et al., 1994). On the basis of punch biopsy analyses
they were able to show, with a high confidence level, that lesions in
which the vessel depth was within 830 μm from the surface of the skin

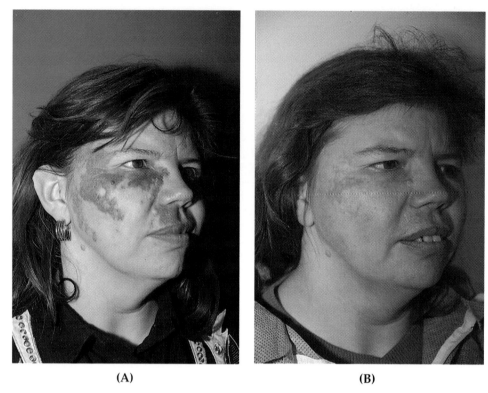

(A)   (B)

Figure 11.10
(A, B) An adult with a grade IV venular malformation before and after treatment
with a copper vapor laser and a flashlamp pumped-dye laser. The lesion responded well and almost completely cleared; however, postinflammatory hyperpigmentation persisted for 6 months before clearing spontaneously.

responded well, whereas lesions in which the mean vessel depth was
greater than 1,000 μm responded poorly or not at all. Given the limited
depth of penetration of yellow light, it seems logical that some lesions
failed to respond completely due to the fact that their vessels are beyond the reach of conventional yellow light lasers. However, this explanation does not satisfy all of the issues. Some lesions may fail to respond on the basis of their vessel diameter or the rate of blood flow
through these lesions. After much speculation, these findings offer a
logical explanation for the difference in response rates.

Certain anatomical sites appear to respond better than others
(Renfro and Geronemus, 1993; Orten et al., 1996). Lesions involving the
forehead, temple, lateral aspect of the face, neck, chest, and shoulders

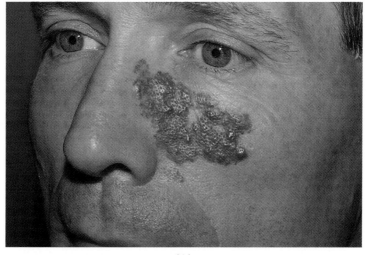

(A)

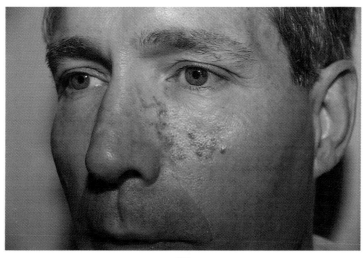

(B)

Figure 11.11
(A, B) An adult with a grade IV venular malformation and significant cobblestone formation before and after treatment with a copper vapor laser and a flashlamp pumped-dye laser. The cobblestones persisted and proved to be resilient to treatment. Because these are made up of much larger vessels, an Nd:YAG laser would have been more appropriate.

required the least number of treatments to achieve a maximal response. These differences are likely due to the differences in the thickness of the skin and the depth of the vessels within these anatomical locations. In addition, lesions that were not confluent or "geographic" in their distribution required fewer treatments than those that were confluent (Orten et al., 1996).

A disturbing finding reported by Orten et al. (1996) concerns the possible recurrence of lesions. During the follow-up evaluation, it became apparent that some patients who had initially responded well experienced some return of their lesion (Fig. 11.12). The number of patients who experienced recurrence was directly related to the time that had elapsed since their last treatment. Forty percent of patients whose last treatment was between 2 and 3 years prior, recurred, and 50% recurred at 4–5 years (although only a small number were followed this long). The degree of recurrence was not noted but it was thought that, given sufficient time, an even greater number would recur. A possible explanation for this lies in the underlying etiology of venular malformations. If a venular malformation is a "sick dermatome" in which there is a relative or a complete absence of autonomic innervation of the superficial pappillary venular plexus, then it stands to reason that whatever blood vessels are left behind after treatment, will recur. The time it takes for a lesion to recur will depend on the grade of the lesion. A high-grade lesion, one in which there is an absolute deficiency of autonomic innervation, will progress and recur more rapidly than a low-grade lesion (one in which there is a relative deficiency of autonomic innervation). This leads to the ultimate question: After having treated a patient several times, and at a considerable cost, what should be done about recurrence? To maintain a good result after maximal clearance, the patient will need a maintenance or "touch up" treatment periodically. The frequency of these will depend on the time it takes for the lesion to recur. Judging from the data on recurrence, this will be between 18 months and 5 years.

With regard to the different response rates claimed by proponents of different laser systems, in all probability the response to laser treatment is constant regardless of the laser used, with the proviso that one use a standard yellow light laser. The response is rather a function of the depth of the vessels that make up the lesion, which is in turn related to the anatomical location of the lesion, not the laser being used. The likely difference between lasers relates more to their respective

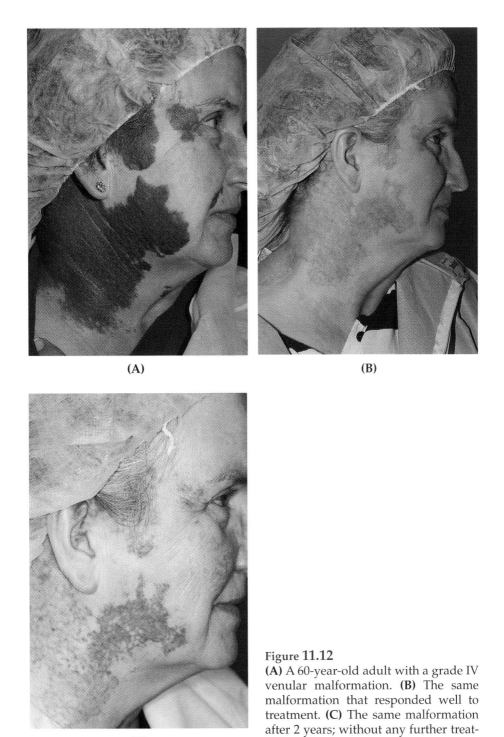

(A)

(B)

(C)

Figure **11.12**
(A) A 60-year-old adult with a grade IV venular malformation. **(B)** The same malformation that responded well to treatment. **(C)** The same malformation after 2 years; without any further treatment, the lesion began to recur.

complication rate and whether they are able to remove the lesion without scarring.

11.3. VENOUS MALFORMATIONS

Several modalities have a role in the management of venous malformations: lasers, surgical excision, and sclerotherapy. A particular lesion may require treatment with more than one modality. The choice of treatment should therefore be individualized and will depend on several factors: the *depth* of the lesion, its *extent*, and its *anatomical location*.

In general, lasers have a limited depth of penetration and are therefore useful for the treatment of superficial lesions. Because the size of the vessels is consistent with our definition of "large vessel disorders," the Nd:YAG laser forms the mainstay of laser treatment. These lasers can only be used in superficial lesions or where there is a superficial component of a compound lesion. In the latter instance, laser photocoagulation will diminish the vascularity of the overlying skin or mucosa, which can then be preserved during surgical resection of the deeper component (Fig. 11.1; see also Chapter 10). Sclerotherapy has also evolved into a viable alternative and can be used alone or as an adjunct to either laser photocoagulation or surgical excision (see Chapter 10). The only modality capable of complete ablation is surgical excision, but this may not be possible with all lesions. Because venous malformations more commonly involve certain sites, a description of the management of lesions at these sites will be used to illustrate the management options.

11.3.1. Venous Malformations of the Lips

Venous malformations commonly involve the upper or lower lip. These lesions usually involve the vermilion and extend to and often beyond the vermiliocutaneous junction. They may also involve the adjacent mucosa and occasionally the buccal fat space (Fig. 11.13). The depth and the thickness of the lesion are extremely variable. Unlike venular malformations, venous malformations have their origins in the subcutaneous or the submucosal plexus. They therefore almost always involve the entire submucosal space and frequently the orbicularis oris muscle as well as the subcutaneous tissues.

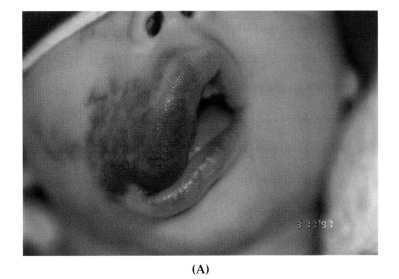

(A)

(B)

Figure **11.13**
(A) An 18-month-old child with a venous malformation involving the full-thick-ness of her upper lip as well as her vermilion. The lesion also extends to her buccal fat space. **(B)** The child has undergone two Nd:YAG laser treatments, a major re-section of the venous malformation, and a wedge resection including a segment of orbicularis oris to reduce the length of her upper lip.

When dealing with a compound lesion, treatment should include prior laser photocoagulation followed with surgical resection. The role of laser photocoagulation is to ablate the superficial component of the malformation (i.e., the mucosal or the cutaneous surface). Subsequent surgical resection can then spare these surfaces, and a mucosal or a cutaneous flap of at least 3–4 mm in thickness can then be raised to expose the underlying lesion. This flap is fibrous and is fairly easy to separate from the underlying venous malformation, a feat that could not have been accomplished without prior laser treatment. Although recurrence is inevitable (unless the entire lesion has been removed), the speed with which this occurs will depend on the grade of the lesion and the extent of the residuum. High-grade lesions will obviously recur more rapidly, and the grade of the lesion can be determined by its presurgical behavior. Surgical resection will need to be complete to prevent recurrence, and this may, of necessity, include muscle as well as the subcutaneous tissues. Preoperative magnetic resonance imaging (MRI) will delineate the full extent of the venous malformation and, in particular, the degree of muscular involvement. It is therefore a mandatory part of the preoperative workup and is especially important because resection of a large portion of orbicularis oris muscle may be necessary and this in turn may give rise to an adynamic segment (Fig. 11.14). The patient or the child's parents should therefore be warned of this possibility prior to surgery if it is indeed necessary to resect muscle. Another consequence of surgery may be a need for a "filler" because removal of the submucosa and/or musclaris layer will leave an extremely thin lip. Dermal allografts will serve this purpose and can be harvested, or, alternatively, lyophilized cadaveric dermis (Alloderm) can be used.

Surgical resection is extremely difficult. The malformation is made up of a sponge of thin-walled ectatic venous channels that bleed easily and are difficult to control. Monopolar and bipolar electrocautery are usually ineffective, especially with the larger vessels, and one may have to resort to hemostatic clips or sutures. Hemostatic agents such as prothrombin or micofibrillar collagen are useful particularly after the lesion has been removed and, with a degree coaxing, will secure hemostasis.

With these techniques, it should be possible to remove most if not all of the lesion without compromising the functional or aesthetic result. The normal vermilion color should return, and the thickness of the lip and the

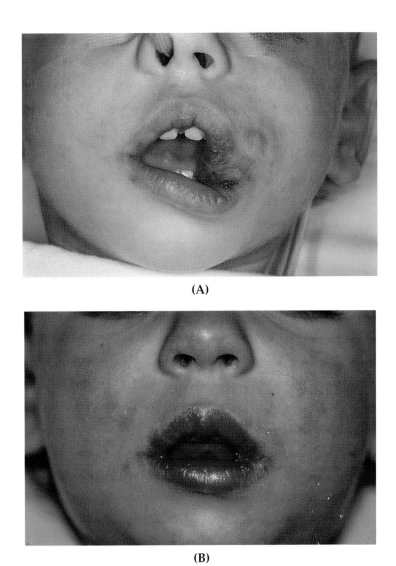

(A)

(B)

Figure **11.14**
(A, B) An 18-month-old child with a venous malformation of her oral commissure
and upper lip before and after treatment with an Nd:YAG laser and surgical resection. The entire lesion was removed through a mucosal incision.

adjacent tissue will be corrected (Figs. 11.14, 11.15). Recurrent venous malformation within the mucosal flap and/or cutaneous tissues can be addressed with further resection or laser photocoagulation at a later stage.
With a cooperative patient, laser treatment can be done under local anesthesia. Residual venous malformations within a deeper plane will manifest much later and will need to be dealt with surgically.

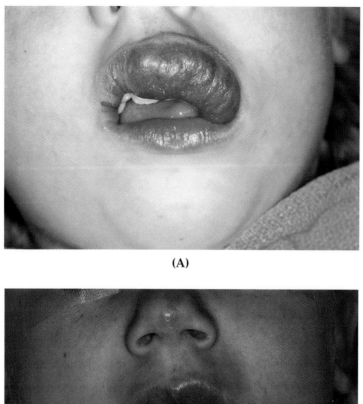

(A)

(B)

Figure 11.15
(A, B) A child with a venous malformation of her upper lip before and after treatment with an Nd:YAG laser followed with surgical resection of the lesion. A mucosal incision and elevation of a mucosal/cutaneous flap to the vermiliocutaneous junction enabled us to remove the malformation, which was external to the orbicularis oris muscle.

11.3.2. Venous Malformations of the Buccal Fat Space

The buccal fat space is an anatomical space that is responsible for the normal contour of the cheek. This space is made up of buccal fat, blood vessels, and the zygomatic and buccal branches of the facial nerve. The space extends deep to the zygoma, where it is limited by temporalis fascia laterally and the temporalis muscle medially. The atomical relations of this space are (Fig. 11.16):

- The masseter muscle and the ramus of the mandible *posteriorly*
- The body of the mandible *inferiorly*
- The nasolabial fold, the upper lip, the oral commissure, and the lower lip *anteriorly*
- The zygoma and the temporal extension *superiorly*
- The space extends from the subcutaneous tissues *laterally* to the buccal mucosa *medially*

The buccal fat space is a common site for venous malformations. In an analysis of 60 patients, this site was second only to the upper and/or lower lips (Waner, unpublished data). Given the natural history

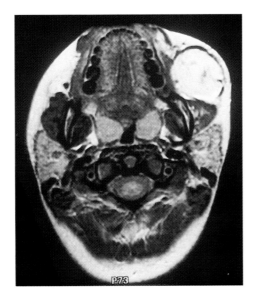

Figure **11.16**
An MRI showing involvement of the fat space with a venous malformation.

of these lesions, one can expect progressive enlargement of the malformation; with time and within the confines of the buccal fat space, this will give rise to progressive facial assymetry (Fig. 11.17). Surgical excision of these lesions can be a challenge, the first of which is the surgical approach. An intraoral approach is tempting but is very likely to fail because exposure is limited and the facial nerve branches are very difficult to identify from this approach. Furthermore, removal of the lesion is at best subtotal since it is impossible to access the entire buccal fat space adequately in this way. This should therefore be discouraged.

Our preferred approach is through an extended parotidectomy incision. The upper limb of the excision should extend superiorily to 3 cm above the tragus. From there, the excision should arch forward to just behind the hairline. The inferior limb should follow a cervical crease about 4 cm below the ramus of the mandible and extend anteriorly to the level of the oral commissure (Fig. 11.18). With these extensions, it will be possible to extend the flap as far forward as the nasolabial line

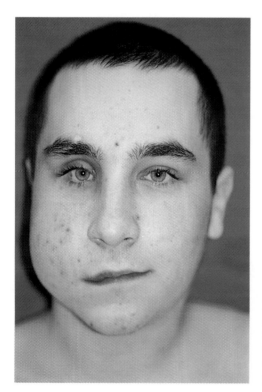

Figure 11.17
An 18-year-old man with a venous malformation of his right buccal fat space. The lesion has progressively enlarged since childhood, deforming his cheek.

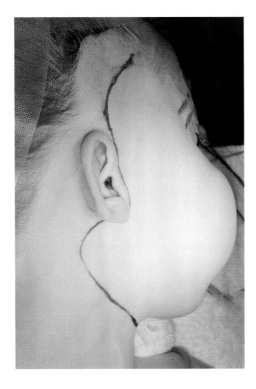

Figure 11.18
A buccal fat space infiltrated with a venous malformation. The buccal and zygomatic branches of the facial nerve can be found in the superficial aspect of this space.

(the anterior extent of the buccal fat space). The flap should be developed superficial to the parotid fascia and deep to the superficial muscular aponeurosis system and extend as far forward as the anterior border of the parotid gland. It is then best to locate the facial nerve in the usual fashion and trace it and its branches through the parotid gland to its anterior border. It is not necessary to remove any of parotid tissue unless the gland is involved. The surgeon may on the other hand elect to locate the buccal and zygomatic branches of the facial nerve along the anterior border of the parotid gland.

Once the location of the facial nerve has been accomplished and traced forward to the anterior border of the parotid, the remaining extent of the flap should be cautiously elevated to its anterior extent, the oral commisure and the nasolabial fold. The surgeon should proceed in a subcutaneous flap plane to avoid the terminal branches of the buccal and zygomatic nerves. As the dissection proceeds several buccal branches will become evident. A nerve stimulator is useful in determining which of these are the dominant branches and their relative functions. As the flap is elevated over the buccal fat space, the venous malformation will become obvious (Fig. 11.19). Any of the superficial extensions of the malformation may be encountered during this maneuver, and it may be difficult to leave these attached to the main specimen. If this is so, these extensions should be cauterized using a standard bipolar electrocautery device.

Once the anterior extent of the flap has been reached, the branches of the facial nerve should be dissected through the buccal fat space and then mobilized (Fig. 11.20). It should only be necessary to trace the buccal and zygomatic branches in this way because the frontal and marginal mandibular branches are outside the surgical field. A facial monitor and a stimulator are extremely useful during this phase of the procedure. It may occasionally become necessary to sacrifice an interconnecting branch or even a minor branch of the nerve to gain adequate exposure of the lesion. Stimulation of the main branches will determine the feasibility of this maneuver, and, provided that the major branches are left intact, this is usually safe. Once the branches have been mobilized, the venous malformation can then be removed. This

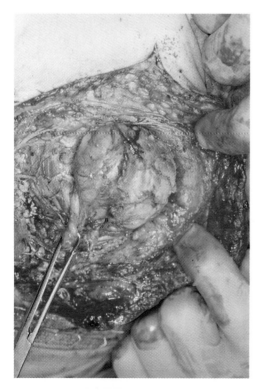

Figure 11.19
The branches of the facial nerve dissected out and mobilized over the venous malformation of the buccal fat space.

Figure 11.20
The venous malformation has been totally removed, leaving a void between the buccal mucosa and skin that will have to be filled with an autologous fat graft.

will usually require dissection over the lateral and anterior surfaces of the masseter and then down to the buccal mucosa. In many cases, the masseter may also be involved with obvious venous malformation. In these cases it is usually markedly hypertrophied. This will have been obvious on MRI and will necessitate removal of the muscle. Considerable hemorrhage can be anticipated because the muscle is extremely vascular. If the entire muscle is to be removed, it should be dissected free from its orgin and insertion, clamped off, and excised. The remaining muscle stumps can then be oversewn to avoid further intraoperative or postoperative hemorrhage.

Amputation of the fat space will leave a void and a subsequent concavity unless the space is filled. An autologous fat graft harvested through a horizontal suprapubic incision will correct this. Slight overcorrection is desirable because the graft will contract by 20%–30% over a 6–18-month period. The wound should then be closed in the normal fashion. Maturation of the graft will result in a soft fat space surrounded by a delicate fibrous capsule. The aesthetic result of this procedure is excellent (Fig. 11.21.).

(A)

(B)

Figure **11.21**
(A, B, C) An 18-month-old child with an extensive venous malformation involving her oral commisure and her buccal fat space. The child was treated twice with an Nd:YAG laser followed by surgical resection of her lesion. The child was 18 months status postresection and was about to undergo a commisurotomy and placement of a "filler" in the depressed area under her scar just lateral to the commisure scar.

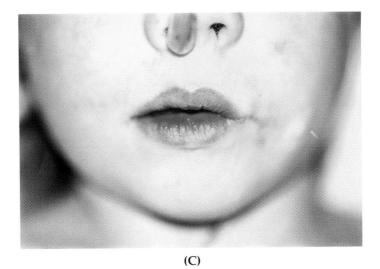

(C)

Figure **11.21** *continued*

11.4. ARTERIOVENOUS MALFORMATIONS

Slow but relentless expansion is the norm, and, as is the case with all other vascular malformations, high-grade lesions present earlier in life and expand more rapidly (Fig. 11.22), whereas low-grade lesions present much later and are more apt to expand slowly (Fig. 11.23). Eventually, gross anatomical distortion, flow reversal with ulceration of the overlying skin, and intermittent hemorrhage will hasten the quest for treatment (Fig. 11.24). Unfortunately, as is often the case, intervention at this late stage will necessitate the sacrifice of a considerable amount of tissue. Timely intervention will not only prevent these complications, but the extent of the resection will also be significantly reduced. Apart from the rare exception, intervention should therefore be planned as soon as the diagnosis has been made. A good working knowledge of the classification of vascular lesions is thus mandatory because the most frequent cause of delayed treatment is misdiagnosis (Fig. 11.24).

Treatment should result in complete eradication of the nidus because this is the fundamental abnormality. The dilated veins and hypertrophic arteries are merely secondary to shunting across the nidus and need not necessarily be resected, although it is difficult to distinguish these from

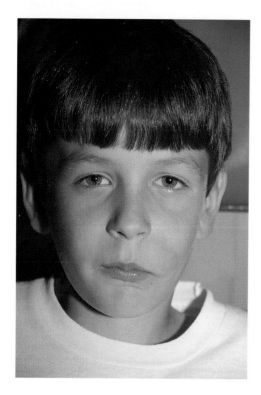

Figure 11.22
A 6-year-old boy with an arteriovenous malformation of his nasolabial fold. This lesion presented soon after birth and recurred soon after incomplete resection. This is therefore a high-grade lesion.

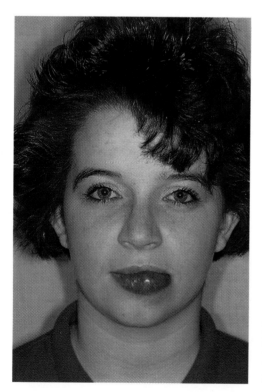

Figure 11.23
A 30-year-old woman with an arteriovenous malformation of her lower lip. This lesion presented for the first time during pregnancy and is therefore a low-grade lesion.

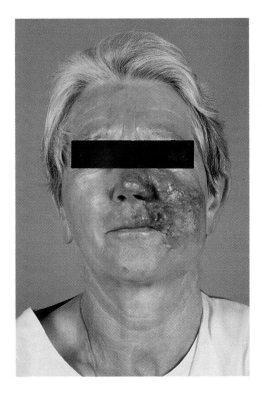

Figure 11.24
A patient with an advanced arteriovenous malformation. Note the ulceration of the overlying skin and the extensive involvement of the soft tissues of the cheek. Because complete resection of this lesion is difficult, and is likely to be destructive, an attempt to control the lesion with palliative embolization is appropriate. Should this fail, an extensive resection will be life saving.

the nidus in an advanced case. If in doubt, the surgeon should err on the aggressive side because even the smallest residual nidus will expand to form a recurrence. Unfortunately, arteriovenous malformations are often more extensive than what is clinically apparent, and incomplete resection is thus common (Fig. 11.25). Appropriate preoperative investigation and planning should prevent this.

11.4.1. Preoperative Planning

The diagnosis of an arteriovenous malformation will be obvious once an MRI and magnetic resonance angiography have been performed (see Chapter 5). A high-flow lesion (with flow voids) and an absence of a parenchymal component will distinguish it from other vascular lesions. In the presence of a low-grade lesion this may not be obvious, and digital subtraction angiography (DSA) may then be diagnostic. Although not usually required for the diagnosis of an arteriovenous malformation, DSA should, in any event, form an integral part of preoperative planning

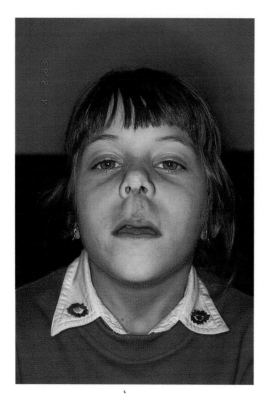

Figure 11.25
A child with a recalcitrant arteriovenous malformation of her upper lip. This lesion had recurred
within weeks of two incomplete attempts at resection. The presence of
a vascular stain of the overlying
skin indicates skin involvement.

for several reasons. These include a determination of the dominant as
well as the colateral blood supply and determination of the feasibility
and/or necessity of preoperative embolization (Baker et al., 1998).

The following information is needed for preoperative planning:

· **Extent:** Particular attention must be paid to the depth and the various anatomical planes involved as well as to the extent of the
nidus. Because this is often more extensive than is clinically apparent, prior knowledge will ensure an adequate margin of resection.

· **Blood supply:** When faced with a large lesion, it may be wise to
gain control of these vessels at an early stage during the procedure.

· **Colateral blood supply:** Both active and potential colateral vessels should be noted because this information will become useful
either during preoperative embolization or at the time of surgery.

11.4.2. Embolization

Embolization is useful, especially with larger arteriovenous malformations and can be used either as a sole therapeutic modality or as adjunctive. The details of each of these roles is considered separately:

- **Adjuvant:** As an adjuvant to surgery, preoperative embolization is most appropriate. Once DSA has mapped out the extent of the nidus, its blood supply and its colateral supply, embolization should proceed from distal to proximal (see also Chapter 10). In this way, both the nidus and its blood supply will be ablated. This will reduce the vascularity of the lesion, thereby reducing the potential intraoperative blood loss. Because a colateral blood supply will become established within a short period of time, the timing of surgery is most crucial and should be about 48 hours after embolization. Gelfoam, polyviyl alcohol (PVA), silicone fluid, and isobutyl-2-cyanoacrylate (IBCA) have all been used successfully to embolize vascular lesions. Gelfoam appears to be the most appropriate preoperative embolic agent because it provides temporary occlusion. The risks associated with migration of Gelfoam and central axis damage or permanent ischemic cranial nerve palsy is thereby minimized.

- **Therapeutic:** Embolization can be used both palliatively and curatively, although the likelihood of permanent "cure" has not been established (Fig. 11.24) (Gomes, 1994). Most interventional radiologists recognize this and refer to "control" rather than "cure." In this instance, a gradual staged procedure, proceeding through the nidus from proximal to distal, to include the active as well as any potential arterial supply, is necessary. All embolic material should be delivered under fluoroscopic screening, and any gradual increase in peripheral resistance should be recognized because this will minimize the risk of backflow of embolic material into the internal carotid circulation. To prevent a recurrence, every vascular space within the nidus should be obliterated since any remaining spaces will inevitably parasitize a colateral blood supply. More often, the role of embolization is palliation of an unresectable lesion or one that has bled and is likely to do so again. This procedure may therefore need to be repeated at some future date.

The choice of embolic material is crucial (Latchaw and Gold, 1979). Nonabsorbable PVA or ICBA is used in most centers. At least one investigator claims that a synergistic effect is seen when PVA is combined with gelfoam. Gelfoam decreases the coefficient of friction, which in turn facilitates distal delivery of a greater number of PVC particles.

Embolization is not without risk. An intimal tear may result in the formation of a thrombus, which may in turn lead to neural infarction. Likewise, embolic material may dislodge with the same catastrophic outcome. The possibility of ischemia with permanent damage to a cranial nerve or skin necrosis is thus always present. It is therefore imperative that the interventional radiologist know and understand the blood supply to each of the cranial nerves.

11.4.3. Surgical Resection

As with all other vascular lesions, slow meticulous dissection with particular attention to hemostasis is necessary. A Shaw scalpel or contact Nd:YAG laser will facilitate this (Baker et al., 1998). Accessible major feeding vessels should be temporarily controlled at an early stage but never sacrificed unless they enter the nidus directly. Proximal ligation of these vessels will only complicate or in some cases preclude further embolization at a later stage because colaterals may develop from the internal carotid system. It is important to differentiate between dilated veins draining the arteriovenous malformation, which may revert back to normal after surgery, and dilated but abnormal vessels, which are part of the nidus. The former can be safely left, whereas leaving the latter will result in a recurrence.

The aim of surgery must be complete removal of the nidus (Baker et al., 1998). To accomplish this, it may be necessary to sacrifice any structure that forms part of the nidus, and the most commonly involved structure is muscle (Fig. 11.26). Occasionally skin and mucosa and rarely, bone and cartilage will need to be sacrificed. Special mention should be made of lesions involving skin. Involvement of the overlying skin is usually clinically obvious and is manifest by the presence of a vascular blush or stain. Because the stain forms part of the nidus, failure to remove this will invariably result in a recurrence (Fig. 11.25). In addition, the everpresent possibility of intracranial involvement should always be excluded prior to resection of these lesions. Lastly,

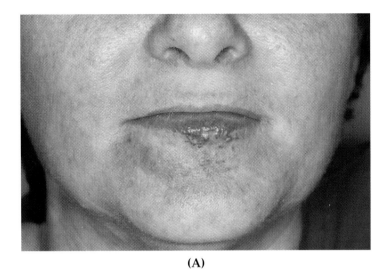

(A)

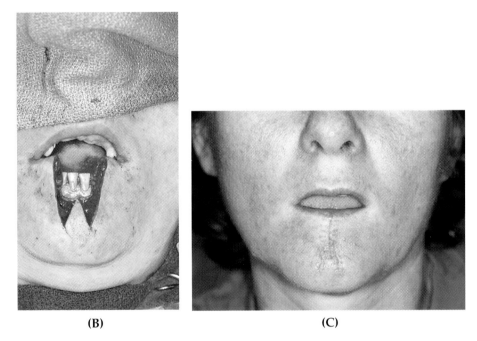

(B)   (C)

Figure 11.26

A 55-year-old woman with an arteriovenous malformation of her lower lip before
(A), immediately after (B), 3 weeks after a (C) wedge resection. Although this was
a low-grade lesion, skin and muscle involvement necessitated resection. A degree
of microstomia was left at the end of the procedure, but this was necessary to
avoid recurrence.

given the high rate of recurrence, the surgical margin should always be generous because the distinction between nidus and related vessels (arterial and venous), may not be that obvious in older, more advanced lesions.

11.5. LYMPHATIC MALFORMATIONS

Of all vascular malformations, lymphatic malformations are the most difficult to erradicate. Their diffuse nature coupled with the difficulty in distinguishing involved tissue from normal tissue make complete surgical extirpation difficult. The likelihood of postsurgical recurrence is thus higher than with other vascular lesions. In general, treatment will vary according to the depth and the extent of the lesion. Superficial mucosal lesions are amenable to ablation with a CO_2 laser, whereas deeper lesions will require surgical excision. In addition, certain other factors need consideration. Microcystic lesions do not respect tissue planes and frequently involve muscle as well as skin or mucosa. Coupled with this is the likelihood of prolonged lymphedema that almost always follows the surgical extirpation of an extensive lesion. Macrocystic lesions, on the other hand, are more likely to be localized and to respect tissue planes, thereby simplifying surgery. They are therefore more amenable to complete excision. This is in contrast with the diffuse microcystic variety, which are usually extensive and will sometimes need to be removed in stages, knowing full well that it may not be possible to completely remove the malformation.

11.5.1. Laser Ablation

CO_2 laser ablation should be reserved for superficial mucosal lesions (Figs. 11.27, 11.28). Because the fluid-filled vesicles are almost always connected with deeper cisterns, ablation of these vesicles alone will usually result in an early recurrence. Ablation should always be continued to a sufficient depth to ensure destruction of most if not all of these cisterns, which may be as deep as the underlying muscle. It is therefore extremely difficult to eradicate the entire lesion, and recurrence is likely. The surgeon should thus use discretion in deciding the appropriate depth to ablate to. In the event of incomplete ablation, a compromise between tissue destruction and destruction of the cisterns will ensure an adequate lapse of time before any recurrence. Hopefully, no

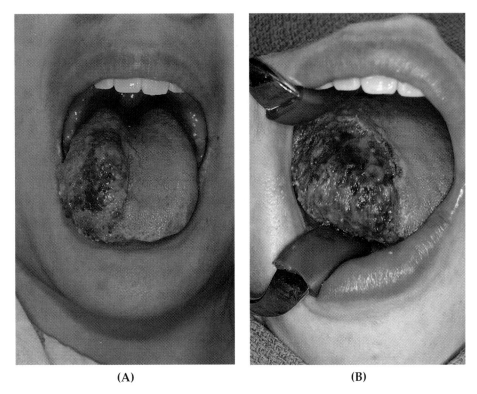

(A)   (B)

Figure 11.27
(A, B) A superficial mucosal lesion of the tongue. Each of the fluid-filled vesicles is connected with a deeper cistern. Laser ablation should therefore be deep enough to destroy these to prevent an early recurrence.

recurrence at all will be seen; however, as is often the case, some form of recurrence is inevitable because it is often impossible to destroy the entire lesion without extensive tissue destruction.

Most patients should be treated under general anesthesia. With an appropriate mouth gag and a tongue stitch, if necessary, adequate exposure should be obtained. A defocused CO_2 laser beam, set at 20 W in the continuous mode, should be used to ablate the lesion in a criss-cross pattern to ensure uniform ablation. As one proceeds through the mucosal vesicles, a honey-combed pattern will be encountered in the submucosal or deeper layers (Fig. 11.27). This represents the cisterns. Ablation should continue in the same uniform manner until no further lymph spaces are encountered or until the surgeon has reached his or her threshold of comfort with the degree of tissue destruction. Frequent removal of the char tissue will assist in this determination.

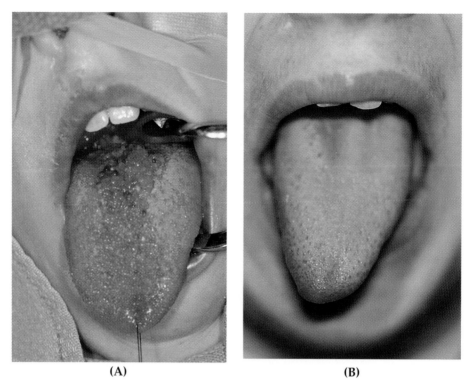

(A)   (B)

Figure 11.28
(A, B) A superficial mucosal lesion before and after treatment with a CO_2 laser.
This patient underwent two laser treatments within a 6-month period. The patient
is now 2 years status postablation and is without evidence of recurrence.

No attempt at closure of the defect should be made. The area will
heal by second intention over the ensuing weeks. Postoperative care
should include frequent mouthwashes with half-strength peroxide and
an appropriate soft diet. Despite the fact that large areas are often
treated, re-epithelialization is usually complete in 2–3 weeks and scarring is usually minimal (Fig. 11.28). A pyogenic granuloma is a frequent
complication and should be ablated in the same way.

11.5.2. Surgical Excision

Most authors agree that surgical resection is the preferred modality
(Hancock et al., 1992; Ogita et al., 1991; Fonkalsrud, 1986). Localized
macrocystic lesions are most amenable to primary excision and rarely

recur. Diffuse microcystic lesions are, as mentioned previously, more difficult problems and may require a staged removal (Fig. 11.29). Care should be exercised in identifying and preserving important cervical and facial structures because anatomical planes are often distorted.

Extensive lesions involving mucosa as well as soft tissue may need to be removed with a combined approach. Surgical excision of the deeper component and laser ablation of the mucosal element will both be necessary but should be separated by at least 6 weeks. We always recommend laser ablation as the first stage, especially if the tongue is involved, because prolonged postsurgical edema will result in tongue protrusion if it has not been dealt with initially (Fig. 11.29). Tongue protrusion is a particularly difficult problem to resolve. Interruption in the lymphatic drainage of the tongue after removal of an extensive lymphatic malformation of the floor of the mouth, even in the absence of direct tongue involvement, will result in lymphedema and, in extreme cases, tongue protrusion. In the absence of direct tongue involvement, the edema may take up to 1 year to resolve, and occasionally a tongue reduction procedure is warranted. In the presence of lingual involvement, this should always be dealt with as completely as possible prior to any extensive mouth floor or upper cervical resection. If in the unfortunate circumstance this order has not been followed, it will be necessary to ablate the lingual involvement as completely as possible. A tongue reduction procedure will only become necessary after enough time has lapsed to allow natural resolution of the edema.

11.5.3. Sclerotherapy

The role of sclerotherapy is still unresolved. Sclerotherapy can be used either as a primary modality or palliatively where the morbidity of surgical resection outweighs the benefits. Bleomycin and OK432 have been used successfully to treat macrocystic lesions with the advantage of the absence of a surgical scar. Unfortunately, multiple treatments are necessary and the results are less impressive with microcystic lesions. Over the course of the next few years, more clinical data and a refinement of techniques will more clearly determine the role of sclerotherapy in the management of lymphatic malformations. These agents are therefore recommended by some as the treatment of choice for localized lesions (Ogita et al., 1991; Yura et al., 1977; Tanigawa et al., 1987; Okada et al.,

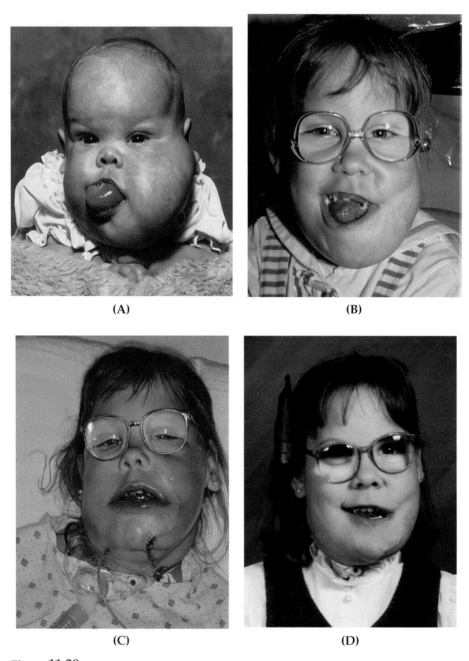

(A)   (B)

(C)   (D)

Figure **11.29**
(A–D) A child with an extensive cervicofacial, microcystic lymphatic malformation. This patient has undergone several procedures, including a debulking procedure of her tongue, a supraglottic laryngectomy to overcome an airway obstruction, a mandibular osteotomy to reduce an open bite, and several debulking
procedures.

1992). The techniques and indications of sclerotherapy are discussed in detail in Chapter 10.

REFERENCES

Achauer, B. M., Vander Kam, V. M., and Padilla, J. F.: Clinical experience with the tunable pulsed dye laser (585nm) in the treatment of capillary vascular malformations. Plast. Reconstr. Surg. 92:1233, 1993.

Baker, L. L., Waner, M., Thomas, J. R., Suen, J. Y., and Bussard, D.: Extracranial arteriovenous malformations of the head and neck. Arch. Otol. Head Neck Surg. (in press), 1998.

Brauner G. J.: Letter to the editor, NEJ Med. 321:902, 1989.

Fonkalsrud, E.: Disorders of the lymphatic system. In Welch, K. J., Randolph, J. G., Ravitch, M. M., et al. (eds.): *Pediatric Surgery.* 4th Ed. Chicago: Year Book Medical, 1986, pp. 1506–1507.

Garden, J. M., Burton, C. S., and Geronemus, R.: Letter to the editor . . . New Engl. J. Med. 321:901, 1989.

Geronemus, R. G.: Pulsed dye laser treatment of vascular lesions in children. J. Dermatol. Surg. Oncol. 19:303, 1993.

Gomes, A. S.: Embolization therapy of congenital arteriovenous malformations: Use of alternate approaches. Radiology 190:191, 1994.

Hancock, B., St. Vil, D., Luks, F., Dilorenzo, M., and Blanchard, H.: Complications of lymphangiomas in children. J. Pediatr. Surg. 27:220–226, 1992.

Kauvar, A. B., and Geronemus, R. G.: Repetitive pulsed dye laser treatments improve persistent portwine stains. Dermatol. Surg. 21:515, 1995.

Latchaw, R. E., and Gold, L. H. A.: Polyvinyl foam embolization of vascular and neoplastic lesions of the head, neck, and spine. Radiology 131:669, 1979.

Ogita, S., Tsuto, T., Deguchi, E., Tokiwa, K., Nagashima, N., and Iwai, N.: OK-432 therapy for unresectable lymphangiomas in children. J. Pediatr. Surg. 26:263–270, 1991.

Oster, J., and Nielson, A.: Nucha naevi and interscapular telangiectasis. Acta Paediat. Scand. 59:416, 1970.

Okada, A., Kubota, A., Fukuzawa, M., Imura, K., and Kamata, S.: Injection of bleomycin as a primary therapy of cystic lymphangioma. J. Pediatr. Surg. 27:440–443, 1992.

Onizuka, K., Tsuneda, K., Ito, S., and Sekine, I.: Efficacy of flashlamp-pumped pulsed dye laser therapy of port wine stains: Clinical assessment and histopathological characteristics. Br. J. Plast. Surg. 1994. Page 271–279.

Orten, S., Waner, M., Flock, S., Roberson, P., and Kincannon, J.: Port-wine stains: An assessment of 5 years of treatment. Arch. Otolaryngol. Head Neck Surg. 122:1174, 1996.

Renfo, L., and Geronemus, R.: Anatomical difference of port-wine stains in response to treatment with the pulsed dye laser. Arch. Dermatol. 129:182–188, 1993.

Reyes, B., Geronemus, R.: Treatment of port-wine stains during childhood with the flashlamp-pumped pulsed dye laser. J. Am. Acad. Dermatol. 23, 1142–1148, 1990.

Rydh, M., Malm, M., Jernbeck, J., and Dalsgaard, C.: Ectatic blood vessels in port-wine stains lack innervation: Possible role in pathogenesis. Plast Reconstr. Surg. 87:419, 1991.

Smoller, B. R., and Rosen, S.: Port-wine stains: A disease of altered neural modulation of blood vessels. Arch. Dermatol. 122:177, 1986.

Tan, O., Sherwood, K., and Gilchrest, B.: Treatment of children with port-wine stains using the flashlamp-pumped tunable dye laser. N. Engl. J. Med. 320:416, 1989.

Tanigawa, N., Shimonatsuya, T., Takahashi, K., et al.: Treatment of cystic hygroma and lymphangioma with the use of bleomycin fat embolization. Cancer, 60:741–749, 1987.

Waner, M.: Classification of port-wine stains. Fac. Plast. Surg. 6:162–166, 1989.

Waner, M., and Suen, J.Y.: The management of congenital vascular lesion of the head and neck. Oncology 9:989–997, 1995.

Yura, J., Hashimoto, T., Tsuruga, N., et al.: Bleomycin treatment for cystic hygroma in children. Arch. Jpn. Chir. 46:607–614, 1977.

Index